PRAISE FOR **MURRAY LONGE**

"Murray distils decades of experience and complex concepts into an accessible roadmap."

KERI A. CLARKE - Start-up industry professional

"Murray has a deep understanding of what he is discussing."

KROL- Amazon Reviewer

PRAISE FOR **MURRAY LONGE'S OTHER BOOKS**

"Like sitting down for an in-depth chat with a knowledgeable, experienced friend in the business. Practical advice based on real world practical success. The writing style makes it easy to read and absorb what might otherwise be quite complex ideas."

WORDSMITH - Amazon Reviewer

"A concise and practical guide for entrepreneurs, packed with clear insights, case studies, and actionable steps. An invaluable resource."

KERI A. CLARKE

"This book is a very well-written, thorough look at every aspect of starting a business. I highly recommend it, especially as it's written by a successful entrepreneur, someone who has actually been through the start-up process from start to finish."

PETER G. - Amazon Reviewer

The Startup Success Series Book 1

Find Problems Worth Solving

How to Validate Your Startup Idea

Murray Longe

WHITMAN
CROSS

Published by Whitman Cross Publishing, Melbourne, Australia.

First edition 2025
Second edition 2026

ISBN 978-1-7645437-6-7 (ebook)
ISBN 978-1-7645437-7-4 (paperback)

NATIONAL LIBRARY OF AUSTRALIA

A catalogue record for this book is available from the National Library of Australia

Disclaimer

The information in this book represents the author's views as of the date of publication. It is for general informational purposes only and is not intended as professional, legal, financial, or business advice. Because circumstances and regulations change, the author reserves the right to revise his views in light of new conditions. Readers should consider their own personal circumstances and consult with qualified professionals before taking any action based on the information in this book. While every effort has been made to ensure the accuracy and completeness of the information, the author and publisher assume no responsibility for any errors, omissions, or inaccuracies. Use of the information is at the reader's own risk.

For more information, visit whitmancross.com

WHITMAN
CROSS

Also by Murray Longe

How To Start Your Start-Up:
The 10 Key Steps To Get From Idea To Investment

Project Management for Startups:
Simple Tools and Templates to Launch, Lead, and Thrive

The Startup Success Series

#1 *Find Problems Worth Solving:*
How to Validate Your Startup Idea

#2 *Beyond The Minimum Viable Product:*
Build Products Customers Really Want

#3 *Scale Without Breaking:*
How to Grow Your Startup Without Losing What Made It Work

#4 *Beyond The Pitch Deck:*
How To Raise Startup Capital

Search All "Murray Longe"

in your amazon store

To every aspiring founder staring at a blank page, wondering if their idea is worth pursuing, this one's for you.

To those who've been told "no" more times than they can count but keep showing up anyway.

To the late-night researchers, the weekend warriors, and the brave souls who dare to ask, "What if?"

This book exists because I believe the startup world needs more voices, more perspectives, and more success stories from people just like you. The knowledge to build something meaningful shouldn't be locked away in expensive courses or exclusive networks. It should be accessible to anyone with the determination to learn and the courage to try.

When I started my entrepreneurial journey, I wished for a guide that would tell me exactly what I needed to know, exactly when I needed to know it. This book, along with my other books, is my attempt to be that guide for you.

Every successful founder started exactly where you are right now - with an idea, a lot of questions, and the audacity to believe they could make a difference.

Your journey begins today. Let's make it count.

Murray Longe
June 2025

Table of Contents

Table of Contents

Table of Contents

Why Most Startups Fail Early

Here's a sobering statistic that should make every aspiring entrepreneur pause: 90% of startups fail. Not because their founders weren't smart, passionate, or hardworking. Not because they lacked funding or technical skills. The number one reason? They built something nobody wanted.

I've been where you are right now. The excitement of a new idea, the conviction that you've spotted something others have missed, the burning desire to jump straight into building mode. But after years of working with startups and building my own businesses, I've learned a crucial truth: falling in love with your solution before validating the problem is the most expensive mistake you can make.

That's why I've written this mini-book—the first in a series designed to give you focused, actionable guidance on specific startup challenges. While my comprehensive guide "How to Start Your Start-Up" provides the full roadmap from idea to investment, this book zeros in on the make-or-break skill that determines whether your startup lives or dies: finding a problem that's worth solving.

What You'll Achieve by Reading This Book

By the time you finish these pages, you'll have:

- A systematic approach to identifying and validating real market problems worth solving
- Proven techniques for conducting customer interviews that reveal genuine insights (not just what people think you want to hear)
- A framework for determining problem-solution fit before you write a single line of code
- The ability to iterate your idea based on real feedback, not assumptions
- Clear warning signs that tell you when to pivot or persist

Most importantly, you'll save yourself months of wasted effort and potentially thousands of dollars by ensuring you're building something people actually need.

Why This Matters Now More Than Ever

In today's fast-moving market, the cost of building has never been lower, but the cost of building the wrong thing has never been higher. With countless startups competing for attention and funding, you can't afford to guess what customers want. You need to know.

Customer discovery isn't just about reducing risk; it's about maximizing opportunity. When you truly understand your customers' problems, you don't just build better products - you build better businesses. You create solutions that sell themselves because they address real, urgent needs that people are already trying to solve.

Your Journey Starts Here

This book distills years of experience and hard-won lessons into a practical guide you can apply immediately. Whether you're a first-time founder with a napkin sketch of an idea or you're three years into building your business and wondering why growth has stalled, the principles in this book will help you build with confidence.

Remember: every successful startup begins with a deep understanding of the customer. Not the technology, not the business model, not even the team - the customer. Master this, and you've mastered the foundation of entrepreneurial success.

Let's dive in and discover what your customers really need.

1

Understanding Market Pain

The Foundation of Every Great Startup

Picture this. You're at a dinner party, and someone starts describing a daily frustration that makes their life harder. As they talk, others around the table start nodding vigorously, sharing their own horror stories about the same problem. The energy in the room shifts. Everyone's engaged, animated, and almost relieved to find others who "get it."

That's market pain in action. And if you want to build a successful startup, you need to become an expert at finding these moments.

Market pain isn't just a mild inconvenience or a "nice-to-have" improvement. It's the grinding, daily friction that makes people actively search for solutions, or cobble together their own workarounds because nothing adequate exists. It's the difference between someone saying "that would be cool" and someone saying "I need this yesterday, and I'll pay whatever it costs."

What Market Pain Really Means

Let me be crystal clear about something: market pain is not what *you* think people need. It's not even what people *say* they need when you first ask them. Real market pain has three critical characteristics:

1. It's Urgent and Frequent. True market pain shows up regularly in people's lives or businesses. It's not a once-a-year annoyance, it's a daily or weekly frustration that compounds over time. The urgency creates emotional energy around solving it.

2. It's Expensive (in Time, Money, or Energy). People are already paying a price for this problem. Maybe they're wasting hours on manual processes, losing money to inefficiency, or burning mental energy on stress and worry. The costlier the current pain, the more valuable your solution becomes.

3. It's Inadequately Solved. This is crucial. Market pain persists because current solutions fall short. Maybe existing options are too expensive, too complex, too limited, or simply don't exist. There's a gap between what people need and what's available.

The Customer Pain Framework

Over the years, I've used a simple framework for categorizing market pain that helps founders think more clearly about the problems they're solving:

Functional Pain

These are practical problems that prevent people from completing tasks efficiently. Think about:

- Time-consuming manual processes that eat up productive hours
- Lack of integration between tools, forcing duplicate data entry
- Missing features in existing solutions that create workflow gaps
- Geographic or accessibility limitations that exclude potential users
- Technical barriers that require expertise most users don't have

Example: Before Dropbox, sharing large files meant burning CDs, using clunky FTP servers, or splitting files across multiple emails. The functional pain was clear: people needed to share files easily across devices and with others. But what made it acute was that this wasn't a once-a-month problem. Designers, developers, and business professionals struggled with this multiple times a day.

Emotional Pain

Never underestimate the power of feelings in driving purchasing decisions. Emotional pain includes:

- Frustration with complexity that makes people feel stupid
- Anxiety about making wrong decisions with serious consequences
- Embarrassment from poor results that affect professional reputation
- Fear of missing out or falling behind competitors
- Stress from feeling out of control or overwhelmed

Example: Warby Parker didn't just solve the functional problem of expensive glasses. They addressed the emotional pain of feeling ripped off by the traditional eyewear industry and the anxiety of choosing frames without trying them on at home. But they went deeper. They understood the embarrassment people felt about not being able to afford designer frames and built a brand that made affordable glasses feel premium.

Social Pain

Humans are social creatures, and many pains stem from how we relate to others:

- Difficulty collaborating with teams across time zones or tools
- Inability to meet social expectations in professional settings
- Challenges in building or maintaining relationships at scale
- Status anxiety in professional or personal contexts
- Communication breakdowns that damage relationships

Example: Slack exploded because it solved the social pain of fragmented team communication. Email chains were killing collaboration, and teams felt disconnected despite working on the same projects. But the deeper insight was that people felt professionally isolated; they were missing the "water cooler" moments that build team culture.

Financial Pain

Money problems create some of the most acute market pains:

- Overpaying for solutions relative to the value received
- Hidden costs and complex fee structures that destroy budgets
- Inability to access financial services due to requirements
- Cash flow challenges that threaten business survival
- Lost revenue opportunities from missing capabilities

Example: Square identified that small businesses were losing sales because they couldn't accept credit cards. The financial pain was direct and measurable - lost revenue. But Square understood something deeper. These business owners felt excluded from the modern economy, unable to serve customers who expected card payment options.

The Pain Intensity Scale: Measuring What Matters

Not all pain is created equal. I use the Pain Intensity Scale to help evaluate whether a problem is *worth* pursuing:

Level 1: Mosquito Bite (Irritating but Ignorable)

- Frequency: Occasional
- Cost: Less than $100/month or 1 hour/week
- Current solution: Living with it
- Emotional response: "It's annoying"
- Purchasing behavior: Will try free solutions only

Level 2: Toothache (Persistent and Distracting)

- Frequency: Weekly
- Cost: $100-1,000/month or 5+ hours/week
- Current solution: Cobbled-together workarounds
- Emotional response: "I really need to fix this"

- Purchasing behavior: Actively comparing solutions

Level 3: Broken Bone (Urgent and Debilitating)

- Frequency: Daily
- Cost: $1,000+/month or 20+ hours/week
- Current solution: Expensive/complex band-aids
- Emotional response: "I'll pay anything to solve this"
- Purchasing behavior: Ready to buy immediately

Level 4: Heart Attack (Mission-Critical Emergency)

- Frequency: Constant
- Cost: Threatens business/life stability
- Current solution: Nothing adequate exists
- Emotional response: "Fix this or we're done"
- Purchasing behavior: Price is secondary to solution

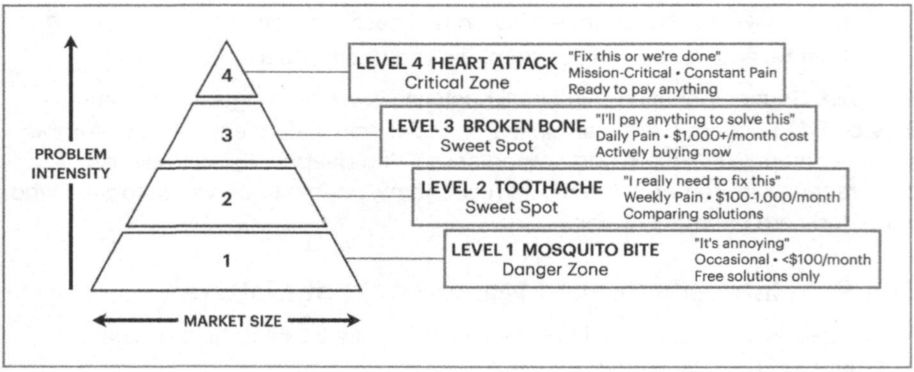

The Pain Intensity Pyramid

Your sweet spot? Levels 2 and 3. Level 1 rarely justifies the switching cost for users. Level 4 is often too niche or requires enterprise-level solutions that are beyond the capabilities of most startups.

Finding Real Market Pain: A Systematic Approach

Now that you understand what market pain looks like, let's talk about how to find it. Too many founders start with a solution in mind and then go looking for a problem to match. That's backward. Here's how to do it right:

Start With Your Own Experience (But Don't Stop There)

The best problems to solve are often ones you've experienced yourself. But, and this is critical, your experience is just the starting point, not validation. Ask yourself:

- What tasks do I dread in my daily work or life?
- Where do I consistently waste time or money?
- What makes me think "there has to be a better way"?

Document these pain points, but don't stop there.

The Experience Audit Exercise:

1. Track your frustrations for one week
2. Note every time you think "there has to be a better way"
3. Rate each frustration on the Pain Intensity Scale
4. Document your current workaround and its cost
5. Ask: "Would I pay to solve this?"

But here's the crucial next step most founders miss: you must validate that others share this pain AND that they experience it the same way you do. Your perception of the problem might be colored by your unique situation, expertise, or resources.

The Observation Method: Becoming a Pain Detective

Some of the best insights come from simply watching people work. Steve Jobs famously said, "It's really hard to design products by focus groups. A lot of times, people don't know what they want until you show it to them." While I don't entirely agree with dismissing all customer input, his point about observation is valid.

Try these observation techniques:

- **Shadow Days**: Spend a day with someone in your target market. Watch them work. Note where they struggle, get frustrated, or create workarounds.
- **Process Mapping**: Document how people currently solve the problem you're interested in. Every step, every tool, every handoff. The pain points will jump out at you.
- **Workaround Spotting**: People's jerry-rigged solutions are goldmines of market pain. Pay special attention to the creative solutions people have cobbled together. Look for:
  - Excel spreadsheets used as databases
  - Post-it notes covering monitors with reminders
  - Multiple tools duct-taped together with Zapier

- Manual processes that should be automated
- "Sneakernet" solutions (physically carrying data)

These are goldmines of market pain.

The Five Whys Technique (With a Twist)

Borrowed from Toyota's manufacturing process, the “Five Whys” helps you dig beneath surface-level complaints. But I've added a twist—the "Five Whos"—to ensure you're not just understanding the problem, but also who experiences it most acutely.

Example Deep Dive:

Surface complaint: "I hate managing my team's projects."

Why? "Because I never know who's working on what."
Who has this problem most? Team leads with 5-15 direct reports.

Why? "Because our communication is scattered across email, Slack, and meetings."
Who suffers most? Remote or hybrid teams across time zones.

Why? "Because we don't have a central place to track everything."
Who needs this most? Teams doing collaborative, interdependent work.

Why? "Because existing project management tools are too complex for our needs."
Who is underserved? Small teams who need simplicity over features.

Why? "Because they're designed for large enterprises, not small teams like ours."
Who would pay? Teams spending 5+ hours/week on coordination.

This approach ensures that you're not only identifying pain but also defining your ideal customer profile. Now you've uncovered a specific market pain - small teams need simple project management tools that don't require extensive training or setup.

The Online Research Method: Digital Anthropology

The internet is a treasure trove of market pain, a 24/7 focus group if you know where to look and how to interpret what you find.

Reddit Deep Dive Strategy: Search for subreddits related to your industry or target market. Look for:

- Repeated questions about how to solve specific problems
- Rants and venting posts about frustrations
- "Does anyone know a tool that..." posts

To do this research:

1. Find relevant subreddits (use redditlist.com for discovery)
2. Sort by "Top" posts of all time to find recurring themes
3. Search for keywords like "frustrating," "hate," "wish," "need help"
4. Document problem patterns, not just individual complaints
5. Pay special attention to posts with high engagement

The Amazon Review Method: Read negative reviews of existing solutions. What specific complaints come up repeatedly? What features do people wish existed?

1. Find products in your problem space
2. Filter for 2-star and 3-star reviews (more honest than 1-star rants)
3. Look for patterns in complaints
4. Note what people bought instead
5. Read questions in the Q&A section

Support Forum Analysis: If you're looking at improving on existing solutions, dive into their support forums. The most common issues and feature requests reveal unmet needs.

1. Access public support forums for existing solutions
2. Categorize issues by frequency and severity
3. Note problems marked "won't fix" or "by design"
4. Identify feature requests with the most votes
5. Look for threads where users share workarounds

Social Media 'Listening': Follow hashtags and join groups where your target market hangs out. Pay attention to the challenges people discuss and the questions they ask.

- Twitter: Search for "anyone know how to [task]" or "[product] alternative"
- LinkedIn: Join industry groups and monitor pain point discussions
- Facebook Groups: Often more candid than professional networks
- Industry Slack/Discord: Where professionals vent freely

The Interview Method: Beyond Surface Answers

While observation and research are valuable, nothing beats direct conversation with potential customers. But here's where many founders go wrong—they ask questions that confirm their assumptions rather than uncovering truth.

Good interview questions are:

- Open-ended: "Walk me through how you currently handle X"
- Specific: "Tell me about the last time you encountered this problem"

- Problem-focused: "What's the hardest part about doing X?"
- Non-leading: Avoid "Wouldn't it be great if you could..."

We'll dive deep into interview techniques in Chapter 3, but for now, your job in these conversations is to listen, not pitch.

The Problem Discovery Interview Framework:

1. Opening Questions (Build Rapport)

- "Tell me about your role and typical day"
- "What's the best/worst part of your job?"
- "What tools do you use most frequently?"

2. Problem Exploration (Uncover Pain)

- "Walk me through the last time you did [task]"
- "What's the hardest part about [process]?"
- "If you had a magic wand, what would you fix?"
- "What tasks do you procrastinate on? Why?"

3. Pain Validation (Measure Intensity)

- "How often does this problem occur?"
- "What happens if you don't solve it?"
- "How much time/money does this cost you?"
- "What have you tried to solve this?"

4. Solution Discovery (Understand Requirements)

- "What would a perfect solution look like?"
- "What's prevented you from solving this already?"
- "Who else needs to be involved in a solution?"
- "What would make you switch from current approach?"

Critical Interview Techniques:

The Silence Technique: After someone answers, count to three before responding. People often fill the silence with their most honest thoughts.

The Specific Instance Method: Always ask for specific examples. "Tell me about the last time this happened" yields better insights than "How often does this happen?"

The Budget Question: "If a perfect solution existed, what would it be worth to you?" This separates 'nice-to-have' from 'need-to-have'.

Validating Market Pain

The Pyramid Test

There is a two-step approach to validating the market pain you've identified. Not all market pain is created equal. I use what I call the **Pyramid Test** to evaluate how irritating a problem really is to a customer, and whether they'd be interested in paying for a solution. The Pyramid Test positions this market pain in one of three categories.

Level 1: Vitamin (Nice to Have)

- People say, "That would be cool"
- Solving it makes life slightly better
- Users might try it if it's free
- Low urgency, low willingness to pay

Level 2: Painkiller (Need to Have)

- People actively seek solutions
- Clear before/after improvement
- Users willing to pay reasonable amounts
- Moderate urgency, clear ROI

Level 3: Critical (Must Have)

- The problem is mission-critical
- Current pain is unsustainable
- Users will pay premium prices
- High urgency, buying decision is when, not if

Your goal? Find Level 2 or Level 3 problems. Level 1 problems might seem easier to solve, but they're incredibly hard to build businesses around because the pain isn't acute enough to drive purchasing decisions.

The DEEP Framework

Once you've found a Level 2 or Level 3 problem, I use the DEEP framework to evaluate whether a problem is worth solving:

D - Demand: Is there sufficient market demand?

- How many people have this problem?
- Is the market growing or shrinking?
- Are people actively seeking solutions?

E - Expense: What's the current cost of the problem?

- Quantify time wasted
- Calculate the money lost

- Assess opportunity cost
- Include the emotional toll

E - Existing Solutions: What's the competitive landscape?

- What are people using now?
- Why do current solutions fall short?
- What would make people switch?
- Is there room for 10x improvement?

P - Propensity to Pay: Will people pay for a solution?

- What's their current budget for this problem?
- Who controls the purchasing decision?
- What's the switching cost?
- Is this a vitamin or a painkiller?

Only pursue problems that score high on all of these four dimensions.

Common Pitfalls in Identifying Market Pain

Let me save you from some expensive mistakes I've seen (and made) repeatedly:

The Projection Fallacy: Assuming everyone experiences a problem the same way you do. A developer's workflow pain is different from a designer's, even if they're on the same team. Or just because you're willing to pay $100 a month to solve something doesn't mean your market is.

The Solution Bias: Getting so attached to your solution idea that you start imagining problems to fit it. This is like having a hammer and seeing everything as a nail. Force yourself to spend 80% of your time on problem discovery, 20% on solution ideation.

The Vocal Minority Trap: Mistaking loud complaints from a small group for widespread market pain. Always validate both breadth and depth. Ten people with severe pain might not be as valuable as 1,000 with moderate pain.

The Feature Shopping List: Confusing a list of desired features with actual market pain. Features are just means to an end. Focus on the underlying problem. "I need better reporting" isn't pain, "I can't show ROI to my boss" is.

The Expert Blind Spot: If you're highly experienced in an industry, you might overlook problems that seem obvious to newcomers. Partner with someone who can bring fresh eyes to your market.

The Geographic Assumption: Assuming pain is universal across markets. What's painful in San Francisco might be irrelevant in Singapore. Always validate locally.

From Pain to Opportunity: The Transition

Identifying market pain is just the first step. The magic happens when you can connect that pain to a solution that's not just viable, but remarkable. As you evaluate market pain, ask yourself:

- Is this problem growing or shrinking?
- Are more people experiencing this pain over time?
- Is the current pain sustainable, or will people be forced to find solutions?
- What would solving this problem unlock for people?

Use this Opportunity Assessment:

Growth Indicators:

- Is this problem growing due to trends (remote work, AI, regulation)?
- Are more people experiencing this pain over time?
- Will solving this unlock new possibilities for users?

Timing Factors:

- Why hasn't this been solved before?
- What's changed to make a solution possible now?
- Is the market ready for a new approach?

Solution Potential:

- Can you deliver 10x improvement over current solutions?
- Is there a clear business model?
- Can you build a moat around your solution?

The best market opportunities exist where several trends intersect. Uber succeeded not just because taxis were painful, but because smartphones were ubiquitous, GPS was accurate, and payment systems were digital-ready. And Zoom exploded not just because video conferencing was painful, but because remote work was growing, internet speeds were increasing, and people were becoming more comfortable with video communication.

Your Next Steps: From Theory to Practice

Understanding market pain is the foundation of customer discovery, but knowledge without action is worthless. Here's your homework:

Week 1: Personal Pain Audit

- Document every frustration you experience
- Rate each on the Pain Intensity Scale
- Choose the top three for deeper investigation

Week 2: Observation Safari

- Shadow someone in your target market for a day
- Document their workflow and pain points
- Identify their current workarounds

Week 3: Digital Detective Work

- Spend 10 hours researching your top problem online
- Create a pain point database from your findings
- Identify patterns and validate assumptions

Week 4: Conversation Campaign

- Conduct 5 problem discovery interviews
- Use the interview framework provided
- Synthesize findings into key insights

Remember, every successful startup begins with someone saying, "There has to be a better way." Your job is to find those moments and dig deeper until you uncover pain worth solving.

The journey from identifying market pain to building a successful solution is challenging, but it starts with genuine curiosity about the problems people face. When you approach customer discovery with empathy and rigor, you don't just build better products—you build products that matter.

In the next chapter, we'll look at how to verify these insights using practical market research methods that don't need a marketing degree or a large budget. You'll learn how to size your market, understand competitive dynamics, and assess whether your identified pain genuinely presents a business opportunity.

For now, focus on becoming a student of problems rather than solutions. The best entrepreneurs fall in love with problems worth solving, not solutions looking for problems.

2

Practical Market Research Methods

From Hunches to Hard Data

So you've identified a market pain that keeps people up at night. You've observed the workarounds, heard the complaints, and felt the frustration firsthand. But here's the sobering truth: your gut feeling isn't enough to build a business on. You need data. Real, actionable data that transforms your hunch into a validated opportunity.

Market research sounds intimidating, like something only big corporations with million-dollar budgets can afford. Let me dispel that myth right now. The most effective market research for startups isn't about hiring expensive consultants or commissioning elaborate studies. It's about being scrappy, strategic, and systematic in how you gather and interpret information.

In this chapter, I'll show you exactly how to conduct market research that's both practical and powerful. Methods you can start using today with minimal resources but maximum impact.

The Market Research Mindset

Before we dive into tactics, let's establish the right mindset. Effective market research for startups isn't academic, it's entrepreneurial. You're not trying to write a dissertation; you're trying to make smart decisions quickly with incomplete information.

The 80/20 Rule of Market Research. You can get 80% of the insights you need with 20% of the effort traditional market research requires. The key is knowing which 20% to focus on. That's what this chapter is about. Giving you the high-impact methods that deliver disproportionate value.

Research as Risk Reduction. Every assumption you make about your market is a risk. Market research is simply the process of systematically reducing those risks by replacing assumptions with facts. The goal isn't certainty, it's confidence that you're heading in the right direction.

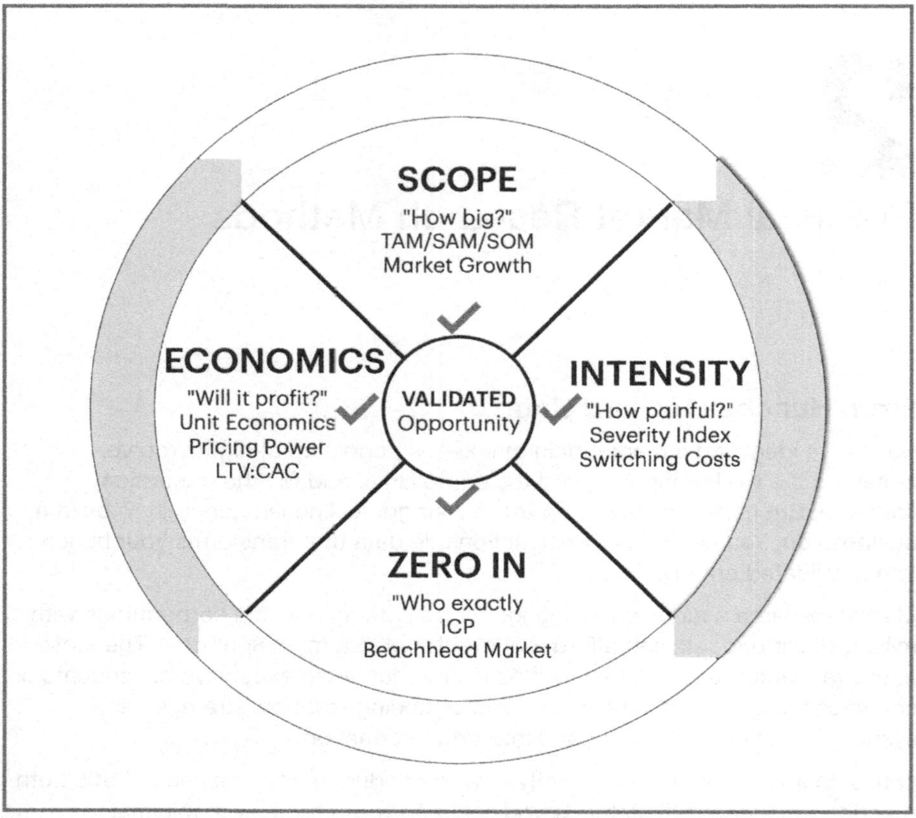

The SIZE Framework Compass

The Market Research Framework: SIZE

I've found a framework called SIZE, an excellent approach to structuring market research for startups. It covers four aspects of the problem and the market:

- **S - Scope**: How big is the opportunity?
- **I - Intensity**: How acute is the problem?
- **Z - Zero In**: Who exactly has this problem?
- **E - Economics**: What's the financial reality?

Let's explore each dimension with practical methods you can implement immediately.

S - Scope: Sizing Your Market Opportunity

The first question every investor will ask is, "How big is the market?" But market sizing isn't just for fundraising. It's about understanding whether you're fishing in a small pond or an ocean.

The TAM-SAM-SOM Pyramid

You've probably heard these acronyms, but let me make them practical:

- **TAM (Total Addressable Market)**: Everyone who could *theoretically* use your solution.
- **SAM (Serviceable Addressable Market)**: Those you can *realistically* reach.
- **SOM (Serviceable Obtainable Market)**: Your realistic capture of customers in the SAM group in 3-5 years.

Here's how to calculate each without an MBA:

Bottom-Up Market Sizing (The Startup Way)

Forget top-down estimates from analyst reports. Here's how to size your market from the ground up:

1. The Unit Economics Method

Start with one customer and expand:

- How much would one customer pay? (Price)
- How often would they pay? (Frequency)
- How long would they remain a customer? (Lifetime)
- How many similar customers exist? (Market size)

Example: Let's say you're building scheduling software for dental practices.

- Price: $200/month per practice
- Frequency: Monthly subscription
- Lifetime: 3 years average (36 months)
- Customer value: $7,200 lifetime value

Now find the number of potential customers:

- Google "number of dental practices in [country]"
- Check industry associations for member counts
- Use LinkedIn Sales Navigator to search for dental practices
- Cross-reference multiple sources

As an example, if there are 200,000 dental practices in the US:

- TAM = 200,000 × $7,200 = $1.44 billion

- SAM = 50,000 (those using digital tools) × $7,200 = $360 million
- SOM = 2,500 (1.25% market share) × $7,200 = $18 million

2. The Comparables Method

Find similar companies and extrapolate:

- Identify 3-5 companies solving similar problems
- Research their revenue and customer counts
- Adjust for your specific differences
- Use this as a reality check on your calculations

Research Tools for Finding Comparables:

- Crunchbase: Funding and revenue data
- Owler: Competitive intelligence
- SimilarWeb: Traffic and engagement metrics
- LinkedIn: Employee counts (use 100-150k revenue per employee as rough estimate)
- Industry reports: These often mention key players and market shares

3. The Search Volume Method

Use search data as a proxy for market interest:

- Use Google Keyword Planner (free with Google Ads account)
- Search for keywords related to your problem
- Look at monthly search volumes
- Calculate: Searches × Conversion rate × Customer value

Example Keywords for Dental Software:

- "dental practice management software" - 5,000 searches/month
- "dentist scheduling app" - 2,000 searches/month
- "dental patient booking system" - 1,500 searches/month
- Total relevant searches: ~10,000/month

If 1% of searchers become customers at $200/month:

- 100 new customers/month
- $20,000 monthly recurring revenue potential
- $240,000 annual run rate from search alone

Market Growth Analysis

Static market size only tells part of the story. You need to understand market dynamics:

Growth Indicators to Research:

1. **Industry Growth Rate**: Is the overall industry expanding?
2. **Technology Adoption Curves**: Where is your market in adopting digital solutions?
3. **Regulatory Changes**: New laws creating demand?
4. **Demographic Shifts**: Generational changes affecting buying behavior?
5. **Economic Factors**: Recession-proof or economically sensitive?

Where to Find Growth Data:

- Google Trends: See search interest over time
- Industry association reports: Often free for basic data
- Government databases: Census, labor statistics, economic data
- Earnings calls: Public companies discuss market trends
- Trade publications: Industry-specific insights

I - Intensity: Measuring Problem Severity

Market size means nothing if the problem isn't severe enough to drive purchasing decisions. Here's how to measure intensity:

The Problem Severity Index

Create a scoring system based on real customer data:

1. Frequency Score (How often does the problem occur?)

- Daily = 5 points
- Weekly = 4 points
- Monthly = 3 points
- Quarterly = 2 points
- Yearly = 1 point

2. Impact Score (What happens when the problem occurs?)

- Business-critical failure = 5 points
- Major productivity loss = 4 points
- Moderate inconvenience = 3 points
- Minor annoyance = 2 points
- Barely noticeable = 1 point

3. Current Solution Pain (How painful are existing solutions?)

- No solution exists = 5 points
- Solutions are prohibitively expensive = 4 points
- Solutions are overly complex = 3 points

- Solutions are adequate but could be improved = 2 points
- Good solutions exist = 1 point

4. Urgency Score (How quickly must it be solved?)

- Immediate (same day) = 5 points
- This week = 4 points
- This month = 3 points
- This quarter = 2 points
- Someday = 1 point

A total score of 16-20 indicates a severe problem worth solving. 12-15 suggests moderate opportunity. Below 12 might not justify the switching costs.

Intensity Research Methods

1. The Day-in-the-Life Study

Document exactly how your target customer experiences the problem:

- Recruit 5-10 target customers
- Have them log every instance of the problem for one week
- Include time spent, money lost, emotional impact
- Calculate the true cost of the problem

Template for Participants:

Date/Time: ____________
Problem Instance: ______________
Time Lost: ______________
Money Impact: _______________
Frustration Level (1-10): ______________
Current Workaround: ____________
Ideal Solution: _______________

2. The Switching Cost Analysis

Understanding what it takes to get customers to change:

- Time to implement a new solution
- Financial cost of switching
- Learning curve for the team
- Risk of disruption
- Political capital required

Research this through interviews: "Tell me about the last time you switched from one solution to another. What was involved?"

3. The Competitive Intensity Map

Create a visual map of how intensely competitors are fighting for this market:

- Number of competitors
- Funding raised by competitors
- Marketing spend (use SEMrush or SpyFu for estimates)
- Pricing pressure
- Customer churn rates

High competitive intensity + High problem severity = Validated market need.

Z - Zero In: Identifying Your Beachhead Market

Trying to serve everyone means serving no one well. You need to identify your beachhead market, the specific segment you'll dominate first.

The ICP Development Process (Ideal Customer Profile)

1. Demographic Segmentation

For B2B:

- Company size (employees and revenue)
- Industry vertical
- Geographic location
- Technology stack
- Growth stage

For B2C:

- Age and life stage
- Income level
- Education
- Location (urban/suburban/rural)
- Lifestyle factors

2. Psychographic Segmentation

- Values and priorities
- Buying behavior
- Technology adoption tendency
- Decision-making process
- Communication preferences

3. Behavioral Segmentation

- Current solutions used
- Buying triggers
- Usage patterns
- Feature priorities
- Price sensitivity

Research Methods for Customer Segmentation

1. The Survey Funnel Method

Design surveys that progressively filter to your ICP:

- Start broad: "Do you experience [problem]?"
- Filter by frequency: "How often?"
- Filter by severity: "How much does this cost you?"
- Filter by willingness to pay: "What would a solution be worth?"
- Collect demographics of those who pass all filters

Survey Tools and Tactics:

- Typeform or Tally: Better completion rates than Google Forms
- Incentives: $10-25 gift cards improve response quality
- Distribution: LinkedIn groups, Reddit, Facebook groups, email lists
- Sample size: Aim for 100+ responses for statistical significance

2. The Cohort Analysis Method

If you have early customers or beta users:

- Segment them by key characteristics
- Analyze which segments have the best metrics:
  - Highest engagement
  - Lowest churn
  - Highest willingness to pay
  - Best word-of-mouth
- Double down on the winning segments

3. The Jobs-to-be-Done Research

Understand not just who buys, but why:

- When do people "hire" your solution?
- What job are they trying to accomplish?
- What's the catalyst for seeking a solution?
- What does success look like to them?

JTBD Interview Script: "Tell me about the last time you [encountered problem]. Walk me through:

- What triggered you to look for a solution?
- What alternatives did you consider?
- What criteria did you use to evaluate options?
- What almost prevented you from moving forward?
- How did you measure success?"

The Beachhead Market Selection Criteria

Choose your initial market based on:

1. **Problem Severity**: Where is the pain most acute?
2. **Accessibility**: Can you reach them easily?
3. **Willingness to Pay**: Do they have budget and authority?
4. **Reference Value**: Will others follow their lead?
5. **Growth Potential**: Can you expand from this base?

E - Economics: Understanding the Financial Reality

Market size and customer profiles mean nothing if the economics don't work. Here's how to research the financial viability:

Price Sensitivity Research

1. The Van Westendorp Price Sensitivity Meter

Ask four questions to find optimal pricing:

- At what price would this be so expensive that you'd never consider it?
- At what price would this be so cheap you'd question its quality?
- At what price would this start to seem expensive, but you'd still consider it?
- At what price would this be a bargain?

Plot the responses to find the optimal price range where you maximize both revenue and perceived value.

2. The Comparative Pricing Analysis

Research how customers currently solve the problem and what they pay:

- Direct competitors' pricing
- Indirect solutions (consultants, manual processes)
- DIY costs (time × hourly rate)
- Cost of not solving the problem

Create a pricing anchor that positions you strategically.

3. The Feature-Value Mapping

Understand which features drive willingness to pay:

- List potential features
- Survey customers: "How much more would you pay for X?"
- Calculate the revenue impact of each feature
- Build your MVP with high-value features only

Business Model Validation Research

1. The Unit Economics Calculator

Build a simple model:

- Customer Acquisition Cost (CAC)
    - Marketing spend per channel
    - Sales team costs
    - Onboarding expenses
- Lifetime Value (LTV)
    - Average revenue per user
    - Gross margin
    - Churn rate
- Target LTV:CAC ratio of 3:1 or better

2. The Sales Cycle Research

Understand how long it takes to close deals:

- Initial contact to first meeting
- First meeting to proposal
- Proposal to decision
- Decision to implementation
- Implementation to first value

Research this by interviewing customers about recent purchases of similar solutions.

3. The Channel Economics Study

Different channels have different economics:

- Direct sales: High touch, high cost, high price
- Self-serve: Low touch, low cost, volume play
- Partner channel: Margin sharing but broader reach
- Marketplace: Fee-based but built-in audience

Research which channels your successful competitors use and why.

Practical Research Tools and Techniques

Free and Low-Cost Research Tools

For Market Data:

- Google Trends: Search interest over time
- Google Scholar: Academic research on your market
- Reddit: Unfiltered customer opinions
- LinkedIn Sales Navigator: (Free trial) Market sizing
- Facebook Audience Insights: Demographic data

For Competitive Intelligence:

- BuiltWith: Technology stack analysis
- SimilarWeb: Traffic and engagement metrics
- Wayback Machine: Historical website changes
- Job postings: Reveal strategic priorities
- Customer reviews: Competitive weaknesses

For Customer Research:

- Calendly: Schedule interviews easily
- Loom: Record interview videos
- Miro/Mural: Visual collaboration for insights
- Airtable: Organize research data
- Typeform: Engaging surveys

The Weekly Research Rhythm

Make research a habit, not a project:

Monday: Competitive Intelligence

- Check competitor websites for changes
- Read new reviews of competitive products
- Monitor their social media activity

Wednesday: Customer Discovery

- Conduct 2-3 customer interviews
- Send follow-up surveys
- Analyze conversation transcripts

Friday: Market Monitoring

- Review industry news
- Check relevant subreddit discussions
- Update your market sizing models

Research Documentation Best Practices

The Research Repository Structure:

```
/Market Research
    /Market Sizing
        /TAM calculations
        /Growth projections
        /Segment analysis
    /Customer Discovery
        /Interview transcripts
        /Survey results
        /Persona documents
    /Competitive Analysis
        /Feature comparisons
        /Pricing studies
        /SWOT analyses
    /Economic Model
        /Unit economics
        /Pricing research
        /Channel analysis
```

The One-Page Research Summary

For each major research initiative, create a one-page summary:

- Research question
- Methodology used
- Key findings (3-5 bullets)
- Implications for our business
- Next steps

This makes insights actionable and shareable with your team.

Common Research Pitfalls and How to Avoid Them

1. Confirmation Bias

- Problem: Only seeing data that supports your assumptions
- Solution: Actively seek disconfirming evidence
- Practice: For every supporting data point, find two challenges

2. Analysis Paralysis

- Problem: Endless research without action
- Solution: Set research deadlines and decision criteria
- Practice: Time-box research to 2-4 weeks per major question

3. Vanity Metrics

- Problem: Focusing on impressive but meaningless numbers
- Solution: Tie all metrics to customer value and revenue
- Practice: Always ask "So what?" about every data point

4. Survey Bias

- Problem: Leading questions that skew results
- Solution: Have a neutral party review your questions
- Practice: Test surveys on colleagues first

5. Small Sample Fallacy

- Problem: Drawing broad conclusions from few data points
- Solution: Set minimum sample sizes before starting
- Practice: 30+ survey responses, 10+ interviews per segment

From Research to Action

Research without action is just expensive entertainment. Here's how to turn insights into decisions:

The Research-to-Decision Framework:

1. **Define Success Criteria Before Researching**
   - What would make us proceed?
   - What would make us pivot?
   - What would make us stop?
2. **Set Research Sprints**
   - Week 1-2: Market sizing and competitive analysis
   - Week 3-4: Customer discovery and segmentation
   - Week 5-6: Economic modeling and price testing
3. **Create Decision Triggers**
   - If TAM > $100M and problem severity > 15, proceed
   - If CAC > LTV/3, revisit business model
   - If no clear beachhead market emerges, continue discovery

4. Document Assumptions to Test

- List every assumption in your business model
- Rank by impact and uncertainty
- Design research to test high-impact, high-uncertainty assumptions first

Your Market Research Action Plan

Stop reading and start researching. Here's your four-week plan:

Week 1: Market Scope

- Calculate TAM using three different methods
- Identify 10 comparable companies
- Create market growth projection

Week 2: Problem Intensity

- Survey 50+ potential customers on problem severity
- Conduct day-in-the-life studies with 5 participants
- Calculate the true cost of the problem

Week 3: Customer Segmentation

- Run segmentation survey to 100+ respondents
- Conduct 10 JTBD interviews
- Define your beachhead market

Week 4: Economic Validation

- Test pricing with 20+ potential customers
- Model unit economics for three scenarios
- Research sales cycles and channel economics

Remember: The goal of market research isn't to eliminate all uncertainty—it's to reduce risk to a level where action makes sense. You'll never have perfect information, but with the methods in this chapter, you'll have enough to make intelligent decisions.

In Chapter 3, we'll dive deep into the art and science of customer interviews—how to get beyond what people say they want to discover what they'll actually buy. But first, put this chapter into practice. Your market is waiting to tell you its secrets; you just need to know how to listen.

The difference between successful startups and failed ones often comes down to this: successful founders do the research others skip. They replace assumptions with insights, hunches with data, and hopes with evidence.

Now it's your turn to join their ranks.

3

Mastering Customer Interviews

The Conversation That Changes Everything

Let me tell you about an expensive mistake I made. I spent six months building a "revolutionary" real estate comparison tool based on what I thought customers wanted. I'd decided “I wanted it”, I’d done surveys, analyzed competitors, run social media campaigns to gauge interest, and even had friends tell me it was brilliant. But I'd never actually sat down with real customers and had deep, probing conversations about what they wanted.

The result? A beautiful product that nobody wanted to buy.

That failure taught me something crucial. Customer interviews aren't just another item on your validation checklist. They're the difference between building something people say they want and something they'll actually pay for. The gap between those two things? It's where startups go to die.
In this chapter, I'll show you exactly how to conduct customer interviews that reveal what people really need, not what they think they want, not what sounds good in a survey, but the raw truth that leads to products people can't live without.

Why Most Customer Interviews Fail

Before we dive into what works, let's be honest about why most customer interviews are worthless:

The Politeness Problem: People are nice. They don't want to hurt your feelings. When you excitedly describe your idea, they'll nod and say, "That sounds interesting!" But "interesting" doesn't pay the bills. You need brutal honesty, not polite encouragement.

The Prediction Fallacy: Humans are terrible at predicting their own behavior. Ask someone, "Would you use an app that does X?" and their answer is almost meaningless. They're imagining a perfect version of themselves, not their real behavior patterns.

The Leading Witness Syndrome: When you're passionate about your idea, it's nearly impossible not to lead the witness. "Wouldn't it be great if..." or "Don't you hate when..." are interview killers. You're planting answers instead of discovering the truth.

The Solution Focus Trap: Most founders can't help but pitch their solution. But the moment you start talking about your product, you've lost the opportunity to learn about the problem. You've shifted from discovery to sales mode.

The Customer Interview Mindset Shift

Effective customer interviews require a fundamental mindset shift. You're not there to:

- Validate your brilliant idea
- Pitch your solution
- Prove your market exists
- Collect compliments

You're there to:

- Understand how people currently solve problems
- Uncover the emotions behind their actions
- Discover what they're already paying for
- Learn why existing solutions fail them

Think of yourself as an investigative journalist, not an entrepreneur. Your job is to uncover the story, not write it.

The Pre-Interview Foundation

Great interviews start long before you meet the customer. Here's how to set yourself up for success:

Recruiting the Right Interview Subjects

The Ideal Interview Candidate:

- Has experienced the problem recently (within 90 days)
- Has tried to solve it (shows problem severity)
- Has budget authority or influence
- Represents your target segment
- Is willing to spend 30-45 minutes with you

Where to Find Them:

1. **Your Network** (but beware of bias)
    - LinkedIn connections
    - Alumni networks
    - Professional associations
    - Slack communities
2. **Cold Outreach That Works**
    - LinkedIn: "I'm researching how [role] handles [problem]. Not selling anything—just trying to learn. Would you have 20 minutes for a virtual coffee?"
    - Email: Focus on their expertise, not your idea
    - Response rates: Expect 10-20% for cold outreach
3. **Incentive Strategies**
    - B2B: Offer to share aggregated insights
    - B2C: Gift cards ($25-50 for 30 minutes)
    - Everyone: Genuine appreciation and follow-up

The Magic Number: Aim for 10-15 interviews per customer segment. You'll start seeing patterns by interview 6-7, and by 10, new insights become rare.

Interview Logistics That Maximize Success

Timing:

- 30-45 minutes (respect their time)
- Tuesday-Thursday, 10am-2pm (highest show rates)
- Avoid Mondays and Fridays

Format:

- Video calls > Phone calls > In-person (for most B2B)
- In-person > Video > Phone (for emotional topics)
- Always have a backup communication method

Recording:

- Always ask permission
- Use tools like Zoom, Loom, or Otter.ai
- Take notes even while recording (shows engagement)

The Interview Script That Actually Works

Here's the exact structure I've used for customer discovery interviews:

Phase 1: The Warm-Up (5 minutes)

Build rapport without wasting time:

Opening: "Thanks so much for taking the time. I'm researching how [people in their role/situation] handle [broad problem area]. There are no right or wrong answers—I'm just trying to learn from your experience."

Context Setting: "I'll be taking notes, and with your permission, I'd like to record this just so I don't miss anything important. Nothing you share will be attributed to you personally. Does that work for you?"

Permission to Dig: "Sometimes I might ask you to go deeper into something that seems small but could be important. Is it okay if I occasionally ask for more detail?"

Phase 2: The Journey Map (10 minutes)

Start with their current reality:

The Opening Question: "Can you walk me through the last time you dealt with [problem area]? Start from the beginning—what triggered the need?"

Journey Mapping Probes:

- "What happened next?"
- "How long did that take?"
- "Who else was involved?"
- "What tools did you use?"
- "Where did you get stuck?"

Emotional Excavation:

- "How did you feel at that point?"
- "What was frustrating about that?"
- "What was at stake if you didn't solve this?"

Phase 3: The Problem Deep Dive (15 minutes)

This is where the gold lies:

Current Solution Dissection: "So it sounds like you're currently using [current solution]. Tell me about that..."

- "What's working well with that approach?"
- "What's not working so well?"
- "If you had a magic wand, what would you change?"
- "What happens if you don't solve this problem?"

The Cost Question Sequence:

1. "How much time do you spend on this weekly?"

2. "What could you be doing instead with that time?"
3. "Has this problem ever cost you money directly?"
4. "What opportunities have you missed because of this?"

The Switching Exploration:

- "Have you tried other solutions?"
- "What made you try something new?"
- "Why did you stick with/leave each solution?"
- "What would have to be true for you to switch again?"

Phase 4: The Dream State (10 minutes)

Paint the picture of their ideal world:

The Outcome Question: "Imagine you've solved this problem perfectly. What's different about your day/work/life?"

Success Metrics: "How would you measure whether a solution is working?"

Value Anchoring: "What would it be worth to you to have this problem completely solved?"

Priority Check: "Where does solving this rank among your other priorities? Why?"

Phase 5: The Wrap-Up (5 minutes)

End strong and keep doors open:

The Referral Ask: "This has been incredibly helpful. Who else do you think struggles with this problem that might be willing to share their experience?"

The Follow-Up Permission: "As I learn more, would it be okay if I reached out with a follow-up question or two?"

The Gratitude: "Your insights have been invaluable. Is there anything I can help you with in return?"

The Art of Interview Archaeology

Great interviewers dig beneath the surface. Here are the techniques that separate rookies from pros:

The Power of Silence

After someone answers, count to three before responding. Those three seconds of silence often yield the most honest, unfiltered insights. People rush to fill silence, and what comes out is usually more truthful than their initial response.

The Specific Instance Method

Never accept generalities. Always push for specific examples:

- Bad: "How often does this happen?"
- Good: "When was the last time this happened? Walk me through that specific instance."

The Five Whys in Practice

Keep digging until you hit bedrock:

Customer: "Our reporting process is a nightmare."

You: "Why is it a nightmare?"

Customer: "It takes forever to compile the data."

You: "Why does it take forever?"

Customer: "Because the data lives in five different systems."

You: "Why is it in different systems?"

Customer: "Because each department bought their own tools."

You: "Why did they buy different tools?"

Customer: "Because IT takes six months to approve anything, so departments go rogue."

You: "Why does IT take so long?"

Customer: "Because they're understaffed and overwhelmed with security concerns."

Now you've uncovered the real problem: IT bottlenecks forcing shadow IT purchases.

The Contradiction Method

When you spot inconsistencies, gently probe: "Earlier, you mentioned this wasn't a huge priority, but you also said you spend 10 hours a week on it. Help me understand that..."

The Hypothetical Price Ladder

To understand value, climb the price ladder: "If a perfect solution existed..."

- "Would you pay $10/month for it?" (Yes)
- "What about $50?" (Probably)
- "What about $100?" (Hmm, depends)
- "What about $200?" (That's getting steep)
- "What would it need to do to be worth $200 to you?"

Advanced Interview Techniques

Once you master the basics, these advanced techniques will unlock even deeper insights:

The Day-in-the-Life Immersion

"I'd love to understand your complete workflow. Could you screen-share and show me exactly how you handle this problem today?"

Watch for:

- Where they pause or hesitate
- Workarounds they've normalized
- The actual time things take
- Their emotional reactions to different steps

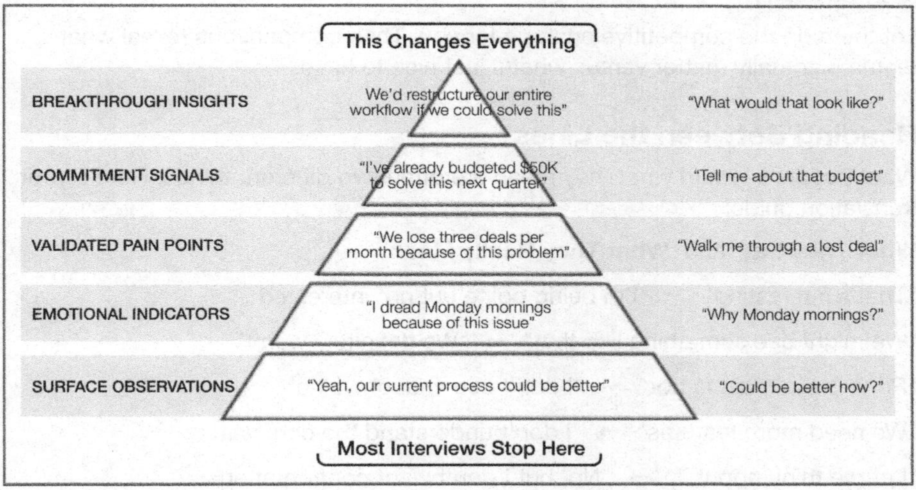

The Interview Insight Pyramid

The Artifact Analysis

"Could you show me any spreadsheets, documents, or tools you've created to deal with this problem?"

People's jerry-rigged solutions reveal:

- What features matter most
- How they think about the problem
- What they're willing to invest effort in
- Gaps in existing solutions

The Buying Process Reconstruction

"Tell me about the last time you bought software/a solution for your team. Walk me through from recognizing the need to making the purchase."

This reveals:

- Who's involved in decisions
- What triggers purchases
- How long buying takes
- What objections arise
- Budget approval processes

The Comparative Analysis

"You mentioned you've used [Competitor X]. How would you compare that to [Competitor Y]?"

Let them do the competitive analysis for you. Their comparisons reveal what features actually matter versus what's just nice to have.

Reading Between the Lines

What people say and what they mean are often two different things. Here's your translation guide:

What They Say and **What They Really Mean**

"That's interesting" → "I'm being polite but not interested"

"We might use something like that" → "We definitely won't"

"Price is the main factor" → "I don't see enough value"

"We need more features" → "I don't understand the core value"

"Let me think about it" → "No, but I don't want confrontation"

"We're happy with our current solution" → "Switching cost is too high"

"Send me more information" → "I want to end this conversation"

Green Light Signals (What You Want to Hear):

"How soon could we start using this?"

"Who else needs to be involved in this decision?"

"We've budgeted for solving this problem"

"I've already tried X, Y, and Z, but none worked because..."

"This is one of our top three priorities"

"I'm spending my own money to fix this"

The Post-Interview Gold Mine

The interview doesn't end when you hang up. The real insights often come from synthesis:

The 24-Hour Rule

Within 24 hours of each interview:

1. Review your notes while your memory is fresh
2. Highlight exact quotes (not paraphrases)
3. Tag insights by category (pain, current solution, budget, etc.)
4. Note what surprised you
5. Write three key takeaways

The Insight Repository

Create a searchable database of insights:

Interview No.:
Date: Participant: (anonymized) Role / Segment:
Key Pain Points:
1. 2.
Current Solution:
What: Cost: $ Satisfaction (1 - 10):
Switching Triggers:
1. 2.
Valuable Quotes:
1. “ “ 2. “ “
Surprising Insights:
1. 2.
Follow-Up Questions:
1. 2.

The Pattern Recognition Process

After every 5 interviews, look for patterns:

- What phrases keep coming up?
- What emotions are universal?
- Where do workflows break down?
- What workarounds are common?
- What price points resonate?

Common Interview Mistakes and How to Avoid Them

Mistake 1: The Pitch Reflex

- Symptom: You start sentences with "What if there was an app that..."
- Fix: Bite your tongue. Count to three. Ask another question about their problem.

Mistake 2: The Happy Path Bias

- Symptom: Only exploring successful outcomes
- Fix: Ask "Tell me about a time when this went wrong"

Mistake 3: The Abstract Discussion

- Symptom: Talking in hypotheticals and generalities
- Fix: Always anchor to specific, recent examples

Mistake 4: The Surface Skim

- Symptom: Accepting the first answers without digging deeper
- Fix: Use "Tell me more about that" liberally

Mistake 5: The Confirmation Shopping

- Symptom: Only hearing what validates your idea
- Fix: Actively seek disconfirming evidence

Turning Interviews Into Action

Great interviews are worthless without action. Here's how to transform insights into decisions:

The Problem Validation Scorecard

Rate each problem on these criteria (1-5 scale):

- Frequency: How often does it occur?
- Severity: How painful is it?
- Spend: Are they already paying to solve it?

- Dissatisfaction: How unhappy with current solutions?
- Accessibility: Can you reach these customers?

Problems scoring 20+ are worth pursuing.

The Solution Requirements Matrix

From your interviews, extract:

- Must-haves (mentioned by 80%+ of potential customers)
- Nice-to-haves (mentioned by 50-79% of potential customers)
- Interesting ideas (mentioned by 20-49% of potential customers)
- Ignore for now (mentioned by <20% of potential customers)

The Go/No-Go Decision Framework

After 10-15 interviews, you should be able to answer:

1. Is this a frequent, painful problem? (If No → pivot)
2. Are people already paying to solve it? (If No → dig deeper)
3. Are current solutions failing them? (If No → find a different angle)
4. Can we deliver 10x improvement? (If No → keep interviewing)
5. Will they pay our target price? (If No → adjust model)

The Continuous Discovery Habit

Customer interviews aren't a one-time validation exercise. The best founders maintain a continuous discovery practice:

Weekly Habits:

- 2-3 customer conversations
- Review and tag insights
- Share learnings with the team

Monthly Rituals:

- Synthesis session to identify patterns
- Update customer personas
- Refine interview questions

Quarterly Deep Dives:

- Interview customers who churned
- Talk to prospects who didn't buy
- Explore adjacent market segments

Your Interview Action Plan

Stop reading. Start talking to customers. Here's your week-by-week plan:

Week 1: Foundation

- Define your target interviewee profile
- Create your interview script
- Recruit 15 potential interviewees
- Practice with 2 internal "safe" interviews

Week 2: Execution

- Conduct 5-7 interviews
- Document insights within 24 hours
- Refine your script based on learnings
- Recruit the next batch of interviewees

Week 3: Synthesis

- Conduct 5-7 more interviews
- Identify emerging patterns
- Create a problem validation scorecard
- Build a solution requirements matrix

Week 4: Decision

- Final 3-5 interviews if needed
- Make a Go/No-Go decision
- Create a presentation of findings
- Plan next steps based on insights

The Meta-Learning From Customer Interviews

After conducting hundreds of customer interviews, here's what I've learned about human nature and startups:

People rarely buy what they say they want. They buy solutions to problems they can't ignore.

The best startup ideas come from listening to problems, not pitching solutions.

Your assumptions are probably wrong. Not because you're not smart, but because you're not your customer.

The founders who succeed are the ones who fall in love with problems, not solutions.

Every great product started with someone saying, "There has to be a better way," and then having the humility to ask, "But what way would actually work for you?"

From Interviews to Iteration

You've identified real pain (Chapter 1), validated market demand (Chapter 2), and now conducted deep customer interviews. But collecting insights is just the beginning. The next chapter will show you how to test problem-solution fit, turning these insights into a solution that customers will actually buy.

But first, go talk to customers. Real ones. Today.

Because here's the truth. Reading about customer interviews is like reading about swimming. You can study the technique, memorize the movements, and understand the theory. But you'll never learn to swim until you jump in the water.

Your customers are waiting to teach you everything you need to know. All you have to do is ask the right questions and listen, really listen, to their answers.

The insights that will make or break your startup are just a conversation away.

4

Problem-Solution Fit Essentials

So far, you've done the hard yards. You've pinpointed a real market problem (Chapter 1), done your research (Chapter 2), and spoken to actual customers who have shared invaluable insights (Chapter 3).

Now comes the moment of truth. Does your solution actually solve their problem in a way that makes them reach for their wallets?

Welcome to the world of Problem-Solution Fit, that magical alignment where what you're building perfectly addresses what your customers desperately need. It's like finding the right key for a stubborn lock, except the lock is your market, and the key is your brilliant idea that's about to change everything (or needs some serious tweaking).

In my main book, *How to Start Your Start-Up*, I introduced the concept of Problem-Solution Fit as part of solving big problems. But achieving Problem-Solution Fit isn't just about having a good idea. It's about having a systematic approach to validate that your solution truly resonates with your target market. And that's exactly what we're diving into in this chapter.

Why Problem-Solution Fit Is Your North Star

Let me be blunt. Without Problem-Solution Fit, you're building a very expensive hobby. I've seen too many founders fall in love with their solution and forget to check if anyone else feels the same way. It's like proposing marriage on the first date. Enthusiasm is great, but you need to make sure the feeling is mutual.

Problem-Solution Fit is the foundation that everything else is built on. You can have the best app, the most innovative technology, or the cleverest business model, but if you're not solving a real problem in a way that customers value, you're just rearranging deck chairs on the Titanic.

Here's why this matters more than anything else in your early-stage journey:

1. **It Validates Your Reason for Existing**: Your startup exists to solve a problem. The Problem-Solution Fit confirms that you're actually doing that, not just thinking you are.
2. **It Saves You Time and Money**: Discovering poor fit early means you can pivot before burning through your savings or your co-founder's patience.
3. **It Attracts Everything Else**: Investors, customers, partners, and talent are all drawn to startups that have clearly demonstrated Problem-Solution Fit. It's like a magnet for resources.
4. **It Guides Your Decisions**: When you have a strong Problem-Solution Fit, every product decision becomes clearer. You can ask, "Does this strengthen our fit?" and actually know the answer.

The Problem-Solution Fit Framework

Let me share a framework I've developed and used with startups. Think of it as your GPS for navigating the treacherous waters of early validation.

PROBLEM-SOLUTION FIT CANVAS	
PROBLEM SIDE	**SOLUTION SIDE**
Customer Segment: Who has this problem?	**Solution Description:** What are you building?
Problem Statement: What specific problem are they facing?	**Key Features:** What capabilities solve the problem?
Root Cause: Why does this problem exist?	**Unique Insight:** What do you understand that others don't?
Current Alternatives: How do they solve it now?	**Why Your Solution:** Why is yours better than alternatives?
Consequences: What happens if unsolved?	**Success Metrics:** How will customers measure success?
	FIT SCORE: ___/10

The Problem-Solution Fit Canvas

This canvas forces you to think critically about both sides of the equation. Too many founders spend 90% of their time only on the solution side and 10% on the problem side. Flip that ratio, and you'll be amazed at how much clearer your path becomes.

The Five Pillars of Problem-Solution Fit

1. Deep Problem Understanding

Remember in Chapter 1 when we talked about identifying market pain? Well, now we need to go deeper. It's not enough to know that people have a problem, you need to understand it better than they do themselves.

The "Five Whys" Technique on Steroids

Here's an enhanced version of the classic "Five Whys". Call it the "Five Whys Plus":

1. Why does this problem exist? (Surface reason)
2. Why is that the case? (Underlying cause)
3. Why hasn't it been solved before? (Market failure)
4. Why do people care about solving it? (Emotional driver)
5. Why now? (Timing and urgency)

Let me give you an example. Say you're building a solution for remote team communication:

1. Why does this problem exist? Because remote teams struggle to stay connected
2. Why is that the case? Because existing tools create information silos
3. Why hasn't it been solved before? Because each tool serves a single purpose, forcing constant context-switching
4. Why do people care about solving it? Because fragmented communication leads to missed deadlines and frustrated employees
5. Why now? Because remote work has become permanent for many companies post-2020

See how we went from a vague problem to a specific, timely, emotionally-charged issue? That's the depth you need.

2. Solution-Problem Alignment

Your solution needs to address the problem like a key fits a lock - precisely and completely. But here's where many founders go wrong. They try to make a Swiss Army knife when customers only want a really good screwdriver.

The Alignment Test

For each feature of your solution, ask:

- Does this directly address a core aspect of the problem?
- Would removing this feature significantly weaken our Problem-Solution Fit?
- Can customers immediately see how this helps them?

If you answer "no" to any of these, you might be 'feature-stuffing' rather than problem-solving.

3. Value Clarity

Customers need to understand your value in seconds, not minutes. If you can't explain how you solve their problem in one sentence, you haven't achieved Problem-Solution Fit yet.

The Mom Test for Solutions

Building on Rob Fitzpatrick's "Mom Test" for customer interviews, here's how to test your solution clarity:

1. Explain your solution to someone unfamiliar with your industry
2. Ask them to explain it back to you
3. If they can't explain it, your Problem-Solution Fit messaging needs work

4. Measurable Impact

A Problem-Solution Fit isn't just a feeling; it's measurable. You need concrete ways to know if your solution is actually solving the problem.

Key Metrics for Problem-Solution Fit:

- **Activation Rate**: What percentage of users who try your solution actually use the problem-solving features?
- **Time to Value**: How quickly do users experience the benefit?
- **Problem Resolution Rate**: What percentage of users report their problem as "solved" or "significantly improved"?
- **Organic Advocacy**: Are users telling others about your solution without prompting?

5. Sustainable Differentiation

Your solution might solve the problem today, but will it still be the best solution tomorrow? This is where your unique insight (I cover this in Chapter 2 of *How to Start Your Start-Up*) becomes crucial.

Quick Validation Tests You Can Run This Week

Theory is great, but you need to test your Problem-Solution Fit in the real world. Here are five validation tests you can implement immediately:

Test 1: The Concierge MVP

Instead of building your full solution, manually deliver the core value to 5-10 customers. Yes, it's unscalable. That's the point. You're testing whether your solution actually solves the problem, not whether it can be scaled.

How to Run It:

1. Find 5-10 people with the problem (from your Chapter 3 interviews)
2. Offer to solve their problem manually using your proposed approach
3. Document everything: time taken, customer satisfaction, actual impact
4. Look for patterns in what works and what doesn't

Test 2: The Feature Fake Door

Create a simple landing page or add a button to your existing product that promises your solution. When users click, tell them it's "coming soon" and gauge interest.

What to Measure:

- Click-through rate on the feature
- Email sign-ups for updates
- Follow-up survey responses about their expectations

Test 3: The Problem-Solution Story Test

Write a one-page story describing a day in the life of your customer before and after using your solution. Share it with potential customers and measure their reaction.

Look for These Responses:

- "This is exactly my life!"
- "I need this now"
- "When can I buy this?"

If you're getting "That's nice" or "Interesting idea," you haven't nailed the Problem-Solution Fit yet.

Test 4: The Wizard of Oz Prototype

Build a simple interface that looks like your solution, but operate it manually behind the scenes. This tests whether users value the outcome without requiring you to build the full technology.

Example: Zappos started by taking photos of shoes in stores, posting them online, and buying them from the store when someone ordered. They validated that people would buy shoes online before building any infrastructure.

Test 5: The Switch Test

Find people currently using alternative solutions (identified in Chapter 2) and offer them yours. The key question: Is your solution compelling enough to make them switch?

Critical Metrics:

- Switching rate: What percentage actually switch?
- Switching friction: What prevents others from switching?
- Retention: Do they stick with your solution or revert?

Common Problem-Solution Fit Mistakes (And How to Avoid Them)

Mistake 1: The Solution in Search of a Problem

You've built something cool, and now you're trying to find a problem it solves. This is like inventing a medicine and then looking for a disease it might cure.

How to Avoid: Always start with the problem. If you've already built something, go back to basics. What problem were you originally trying to solve? Has that changed?

Mistake 2: The Vitamin vs. Painkiller Confusion

Vitamins are nice-to-have; painkillers are must-have. Too many startups think they're building painkillers when they're actually building vitamins.

How to Test: Ask customers: "If this solution disappeared tomorrow, how would that impact you?" Their answer tells you everything.

Mistake 3: The Feature Frankenstein

You keep adding features hoping one will create Problem-Solution Fit. This usually creates a confusing product that solves nothing well.

How to Avoid: Strip down to one core feature that solves one core problem. Nail that before expanding.

Mistake 4: The Echo Chamber Validation

You're only validating with people who think like you or are too polite to give honest feedback.

How to Avoid: Actively seek out skeptics and people who've rejected similar solutions. Their feedback is gold.

Mistake 5: The Premature Scale

You think you have a Problem-Solution Fit because a few people said your idea was good, so you start scaling.

How to Avoid: Use the scoring system below. Don't scale until you consistently score above 20.

The Problem-Solution Fit Scoring System

Here's a simple but powerful way to quantify your Problem-Solution Fit:

PROBLEM-SOLUTION FIT SCORECARD	
	Score
PROBLEM CLARITY (0-2 points each)	**(0-10)**
☐ Customers can articulate the problem clearly	
☐ The problem occurs frequently	
☐ Current solutions are inadequate	
☐ Customers actively seek solutions	
☐ The problem has measurable negative impact	
SOLUTION FIT (0-2 points each)	**(0-10)**
☐ Solution directly addresses root cause	
☐ Customers understand value immediately	
☐ Solution is significantly better than alternatives	
☐ Customers achieve desired outcome	
☐ Solution creates lasting value	
VALIDATION STRENGTH (0-2 points each)	**(0-10)**
☐ Multiple customers confirm fit	
☐ Customers willing to pay	
☐ Early users actively engage	
☐ Organic word-of-mouth occurring	
☐ Retention/repeat usage demonstrated	
TOTAL SCORE: ___/30	
INTERPRETATION:	
25-30: Strong Problem-Solution Fit - Scale confidently	
20-24: Good fit - Refine and strengthen	
15-19: Moderate fit - More validation needed	
Below 15: Weak fit - Consider pivoting	

When to Pivot vs. Persevere

This is the million-dollar question every founder faces. Here's my framework for deciding:

Pivot When:

- After 3 months of testing, your Problem-Solution Fit score is still below 15
- Customers consistently say, "It's interesting but not for me"
- You're solving symptoms, not root causes
- The problem isn't painful enough to drive behavior change
- Your unique insight was wrong

Persevere When:

- Your score is 15+ and improving
- A specific customer segment shows a strong fit (even if others don't)
- Customers are upset when you take the solution away
- The problem is real, but your execution needs work
- You have a clear path to strengthen fit

Connecting Problem-Solution Fit to Your Next Steps

Achieving Problem-Solution Fit is not the end. It's only the beginning. It's your license to move forward with confidence. Here's how it connects to everything else:

1. **It Informs Your MVP**: Once you have Problem-Solution Fit, you know exactly what features to include in your Minimum Viable Product. No guessing, no feature bloat.
2. **It Shapes Your Messaging**: Your validated Problem-Solution Fit becomes your value proposition. You can now clearly articulate why customers should care.
3. **It Attracts Investment**: Investors are 10x more interested in startups that can demonstrate clear Problem-Solution Fit. It de-risks their investment.
4. **It Guides Your Roadmap**: Every future feature can be evaluated against whether it strengthens your Problem-Solution Fit.

Your Problem-Solution Fit Action Plan

Here's what you need to do in the next 7 days:

Day 1-2: Fill out the Problem-Solution Fit Canvas for your startup. Be brutally honest.

Day 3-4: Choose two validation tests from this chapter and set them up.

Day 5-6: Run the tests with at least 10 potential customers.

Day 7: Score your Problem-Solution Fit and decide your next move.

Remember, Problem-Solution Fit isn't a one-time achievement—it's an ongoing relationship you maintain with your market. Markets change, problems evolve, and new solutions emerge. Your job is to stay so closely connected to your customers that your Problem-Solution Fit actually strengthens over time.

The Bottom Line

Problem-Solution Fit is the difference between a startup that struggles and one that soars. It's not about having the most innovative technology or the slickest design—it's about solving a real problem in a way that customers value and will pay for.

As I mentioned in *How to Start Your Start-Up*, great companies are built on solving big problems. But now you have the tools to validate that you're not just solving a big problem—you're solving it in the right way for the right people.

In the next chapter, we'll dive into how to iterate on your idea based on the feedback you've gathered. Because here's the truth: your first attempt at Problem-Solution Fit probably won't be perfect. And that's exactly where the magic happens—in the iterations that transform a good idea into a great business.

So go forth and validate. Your future customers are waiting for the solution only you can provide. Make sure you're building exactly what they need.

5

Iterate Based on Feedback

The Art of Strategic Pivoting

Let me share a story that changed how I think about building startups. A founder I was mentoring had spent months validating a problem, conducting stellar customer interviews, and building what seemed like the perfect solution. Beta users loved it. The feedback was glowing. Then came launch day.

Crickets.

The same people who raved about the product in beta wouldn't pay for it. The founder was devastated. "But they said they loved it!" he kept repeating. That's when I introduced him to a harsh truth. There's a canyon between "love" and "buy," and the only bridge across it is iteration based on real feedback.

That founder went on to iterate his way to a successful exit, but only after learning to read between the lines of feedback and make hard decisions about what to change, what to keep, and what to kill. This chapter will teach you how to master that same skill.

The Iteration Mindset: Your Competitive Advantage

Most founders treat feedback like validation—proof that they're on the right track. But feedback isn't a pat on the back; it's a compass pointing toward product-market fit. The problem is that the compass doesn't point straight north. It spins, wobbles, and sometimes points in opposite directions. Your job is to find true north through intelligent iteration.

The Three Pillars of Effective Iteration:

1. **Humility Over Ego**: Your idea is probably wrong in some important way. That's not failure, that's normal. The failure is refusing to see it.

2. **Speed Over Perfection**: Fast, imperfect iterations beat slow, perfect ones every time. You're not building a monument. You're discovering what works.

3. **Evidence Over Opinion**: Your opinion, your team's opinion, even your customers' opinions, matter less than their actual behavior.

The Feedback Collection Framework

Before you can iterate, you need feedback worth iterating on. Not all feedback is created equal, and collecting the wrong kind is worse than collecting none at all.

The Feedback Hierarchy

The most important feedback (Level 1) down to the least important (Level 4).

Level 1: Behavioral Data (What People Actually Do)

- Usage analytics
- Payment information
- Churn data
- Feature adoption rates
- Time-to-value metrics

Level 2: Contextual Feedback (What People Do and Why)

- Session recordings with user comments
- Support ticket patterns
- Sales call objections
- Cancellation interviews
- Feature request patterns

Level 3: Solicited Feedback (What People Say When Asked)

- NPS surveys
- Customer interviews
- Beta tester feedback
- Feature polls
- Review responses

Level 4: Unsolicited Opinions (What People Volunteer)

- Social media mentions
- Casual comments
- Forum discussions
- Email suggestions

Weight your iteration decisions by this hierarchy. A single piece of behavioral data outweighs a hundred opinions.

Building Your Feedback Engine

1. The Automated Collection System

Set up these tools from day one:

- **Analytics**: Mixpanel, Amplitude, or even Google Analytics
- **Session Recording**: Hotjar or FullStory
- **Error Tracking**: Sentry or Bugsnag
- **Customer Feedback**: Intercom or Drift
- **NPS Tracking**: Delighted or Wootric

2. The Human Collection Process

Create regular touchpoints for qualitative feedback:

- **Weekly Customer Calls**: 3-5 conversations with active users
- **Churn Interviews**: Call everyone who cancels
- **Feature Follow-ups**: Check in 2 weeks after someone uses a new feature
- **Quarterly Business Reviews**: Deep dives with key customers

3. The Synthesis Ritual

Every Friday, spend 2 hours on feedback synthesis:

- Review quantitative metrics
- Read through qualitative feedback
- Identify patterns and outliers
- Create action items for iteration
- Share insights with the team

The Signal vs. Noise Problem

Here's where most founders fail: they treat all feedback equally. But feedback is like a radio - it has lots of static with occasional clear signals. Learning to tune into the right frequency is crucial.

Signals That Matter

High-Value Signals:

- Consistent behavior patterns across multiple users
- Feedback tied to churn or renewal decisions
- Problems preventing users from achieving the core value
- Requests from users who match your ideal customer profile
- Issues affecting willingness to pay or recommend

Example Signal: "Five enterprise customers all mentioned they can't use the product without SSO, and three have delayed purchasing because of it."

Noise to Filter Out:

- Feature requests from non-paying users
- Opinions from people outside your target market
- Suggestions that contradict your core value proposition
- Edge cases affecting <5% of users
- "Wouldn't it be cool if..." ideas

Example Noise: "My cousin tried the app and thinks you should add cryptocurrency payments and a dark mode."

The 10-3-1 Rule

For every 10 pieces of feedback:

- 3 might be signals worth investigating
- 1 will be actionable and impactful

Don't iterate on noise. It dilutes your product and confuses your market.

The Iteration Decision Framework

After gathering feedback, you still need a disciplined way to decide what to build next. I start with the ICE framework, then tweak it for startup reality.

Original or Classic ICE (created by growth-hacker Sean Ellis)

- Score each idea, feature, or project 1–10 (*10 = highest/best/fastest*) on **Impact** (expected benefit), **Confidence** (likelihood the benefit will happen), and **Ease** (speed or simplicity).
- Multiply the three scores: **ICE = Impact × Confidence × Ease**.
- Rank by the total, highest total wins priority, execute the top item, and re-score as circumstances change.

This approach is fast, lightweight, and perfect for agile teams that reprioritise often—but it's subjective and tends to over-reward "easy wins" or the "low-hanging" fruit.

My "Startup-friendly" ICE

I flip the last dimension and the math:

ICE = (Impact × Confidence) ÷ Effort
(Effort is 1–10, where 10 = hard/time-consuming).

This version highlights high-impact, high-confidence work that requires minimal effort, resulting in a more balanced backlog compared to the original multiplication formula.

Why the **Effort-denominator** helps:

1. **It provides a more intuitive discussion of cost:** Teams already talk in story-points or days; dividing by Effort maps cleanly to that language.
2. **There's less "noise" from Ease scores:** Because Effort sits alone in the denominator, a wild mis-estimate there distorts the total less than multiplying all three numbers together.
3. **It's closer to ROI thinking:** You're effectively doing value ÷ cost, which resonates with finance-minded stakeholders.

In short, you still get ICE's speed, but with a clearer value-for-cost lens that surfaces the ideas most worth your limited startup resources.

ICE Scoring for Iterations

When deciding on scores for the three dimensions of ICE, consider these variables below:

Impact (1-10): How much will this improve key metrics?

- Revenue impact
- Retention improvement
- User satisfaction increase
- Market expansion potential

Confidence (1-10): How sure are we that this will work?

- Based on behavioral data (high confidence)
- Multiple customers requesting (medium confidence)
- Theoretical improvement (low confidence)

Effort (1-10): How hard is this to implement? Everything takes some effort.

- Development time
- Opportunity cost
- Risk to existing features
- Ongoing maintenance burden

Formula: Priority Score (ICE Score) = (Impact × Confidence) ÷ Effort

Anything with a Priority Score of 15 or more deserves serious consideration.

The Four Types of Iterations

1. Feature Iterations (Adding/Removing/Modifying Features)

When to do it:

- Clear behavioral data shows feature gaps
- Multiple customers cite a specific feature as a purchase blocker
- Usage data shows features being used in unintended ways

When to avoid it:

- Requested by a vocal minority
- Competes with core value proposition
- Adds complexity without proportional value

Example: Slack didn't start with threaded conversations. They added it only after behavioral data showed messages getting lost in busy channels, and multiple enterprise customers said it was blocking adoption.

2. Positioning Iterations (Changing How You Describe/Market)

When to do it:

- Sales conversations reveal a misunderstanding of value
- Wrong types of customers signing up
- High churn due to mismatched expectations

When to avoid it:

- The team is bored with the current messaging
- Competitor changes their positioning
- One influential customer suggests it

Example: Instagram started as Burbn, a location-based check-in app. User behavior showed people only used the photo-sharing feature, so they repositioned entirely around photo sharing.

3. Pricing Iterations (Adjusting Price/Model/Tiers)

When to do it:

- Conversion data shows price sensitivity
- Customer segments have vastly different willingness to pay
- Usage patterns don't match pricing tiers

When to avoid it:

- Competitors change pricing
- A single customer negotiates hard
- Revenue targets aren't met (without other indicators)

Example: Zoom discovered through usage data that their original per-room pricing model didn't match how customers used the product. They iterated to per-host pricing, dramatically improving adoption.

4. Market Iterations (Changing Target Customer/Use Case)

When to do it:

- Unexpected segment shows superior metrics

- Original market proves too small/competitive
- Adjacent market pulls you in organically

When to avoid it:

- Grass looks greener elsewhere
- Single large customer in new market
- Team wants new challenge

Example: Shopify originally targeted snowboard shops. When other e-commerce businesses started adopting it, they iterated to serve the broader market while maintaining simplicity.

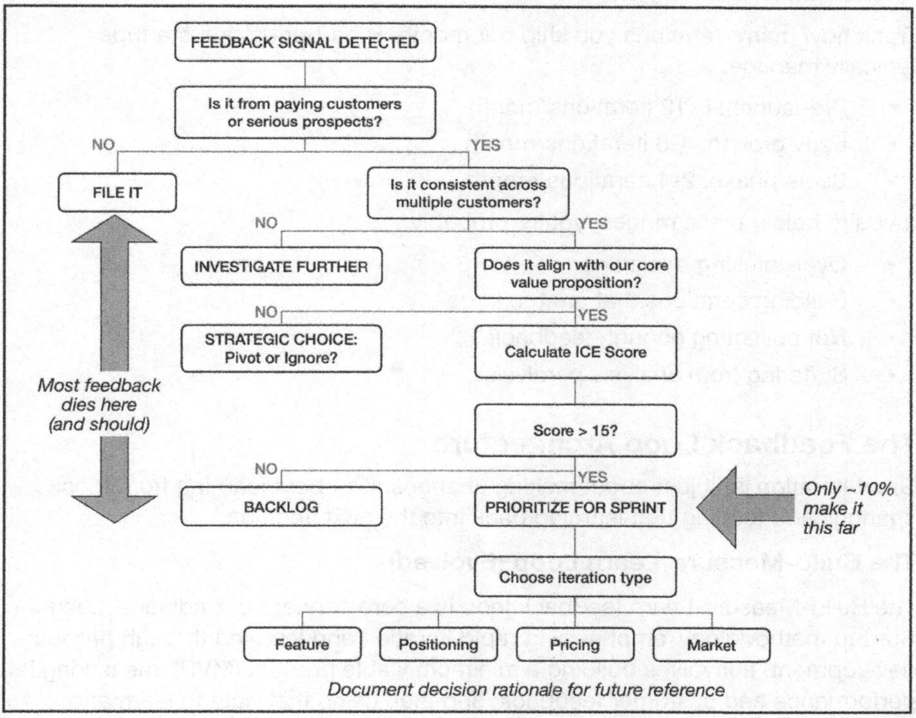

The Iteration Decision Tree

The Art of the Small Bet

The biggest mistake founders make with iteration is going too big too fast. Every iteration should be a small bet that you can learn from quickly.

The Two-Week Rule

No iteration should take longer than two weeks to implement and test. If it does, break it down:

Too Big: "Rebuild onboarding flow based on user feedback"

Right-Sized Iterations:

- Week 1: Add progress indicator to onboarding
- Week 2: Simplify step 3 (where 40% drop off)
- Week 3: A/B test removing step 5
- Week 4: Add helpful tooltips to confusion points

The Iteration Velocity Metric

Track how many iterations you ship per month. High-performing startups typically manage:

- Pre-launch: 8-12 iterations/month
- Early growth: 4-8 iterations/month
- Scale phase: 2-4 iterations/month

If you're below these ranges, you're probably:

- Over-thinking decisions
- Building iterations that are too large
- Not collecting enough feedback
- Suffering from analysis paralysis

The Feedback Loop Architecture

Great iteration isn't just about making changes. It's about learning from those changes and feeding that learning back into the next iteration.

The Build-Measure-Learn Loop (Evolved)

The Build-Measure-Learn feedback loop is a core concept of traditional Lean Startup methodology, emphasizing rapid iteration and learning through product development. It involves building a minimum viable product (MVP), measuring its performance and customer feedback, and then using that data to learn and improve, either by iterating on the product or pivoting to a new approach.

That sounds simple, straightforward, and relatively easy. The reality is messier and more powerful:

Build (1-3 days)

- Ship the smallest viable change
- Include measurement hooks
- Prepare a rollback plan

Release (1 day)

- Soft launch to a subset of users
- Monitor for critical issues
- Gather initial reactions

Measure (3-7 days)

- Collect quantitative data
- Gather qualitative feedback
- Compare to baseline metrics

Analyze (1 day)

- Identify what worked/didn't
- Understand why
- Document surprises

Decide (1 day)

- Double down on success
- Pivot from failure
- Plan next iteration

Share (Ongoing)

- Communicate learnings to the team
- Update documentation
- Inform future iterations

The Iteration Cadence Calendar

Ongoing improvement involves working to a schedule and adhering to it consistently. Each team and organisation will develop a schedule that works for them, but one schedule could look like this:

- **Monday**: Review weekend data and plan the week's iterations.
- **Tuesday-Wednesday**: Build and internal test.
- **Thursday**: Release iterations.
- **Friday**: Initial measurement and analysis.
- **Following Monday**: Decide on next steps

This rhythm creates predictable progress and prevents iteration paralysis.

Learning to Kill Your Darlings

The hardest part of iteration isn't adding—it's subtracting. Features you've built, messaging you've crafted, markets you've targeted. Sometimes the best iteration is elimination. Don't hang on "hoping" something will get better.

The Feature Funeral Framework

When behavioral data shows a feature isn't working, use this approach:

1. **Verify the Signal**
    - Less than 10% of users engage with it
    - It doesn't impact key metrics
    - Support burden exceeds value
    - It complicates the core experience
2. **Run the Sunset Test**
    - Soft disable for 5% of users
    - Monitor for complaints
    - Check the impact on metrics
    - If silence, expand to 100%
3. **Communicate Gracefully**
    - Email affected users (usually very few)
    - Explain why the change
    - Offer alternatives
    - Thank them for the feedback

Example: Google's graveyard of products shows even giants must kill features. The difference? They do it based on data, not emotion.

Advanced Iteration Strategies

Once you master basic iteration, these advanced strategies can accelerate your path to product-market fit:

The Cohort Iteration Method

Instead of iterating for everyone, iterate for specific cohorts:

Step 1: Segment users by behavior, not demographics

- Power users vs. casual users
- Fast activators vs. slow activators
- High value vs. low value

Step 2: Create cohort-specific iterations

- Power user cohort: Advanced features
- New user cohort: Simplified onboarding
- Enterprise cohort: Admin controls

Step 3: Measure impact by cohort

- Did power users become more powerful?
- Did new users activate faster?
- Did enterprises expand usage?

Step 4: Roll out winning iterations to the broader audience

The Competitor Response Framework

When feedback says, "Competitor X does this better":

Don't: Copy their feature directly. **Do**: Understand why customers value it.

Process:

1. Interview 5 customers who mentioned the competitor
2. Understand the job they're trying to do
3. Design your own solution to that job
4. Test if your approach resonates better

Example: When Notion users said "Airtable has better databases," Notion didn't copy Airtable. They created their own database approach that fit their all-in-one workspace philosophy.

The Anti-Iteration: Strategic Stubbornness

Sometimes, the best iteration is no iteration at all. Hold firm when:

Your Core Thesis Is Under Attack: If feedback challenges your fundamental belief about the problem or solution, dig deeper before making a change. You might be ahead of the market.

The Loud Minority Phenomenon: When 5% of users are very vocal about a change but 95% are silent, check the silent majority's behavior before reacting.

The Feature Creep Trap: When iterations are making your product more complex without proportional value, it's time to stop adding and start refining.

Example: Apple famously ignores feedback requesting more customization options. Their strategic stubbornness regarding simplicity is at the core of their value proposition.

Measuring Iteration Success

How do you know if your iterations are working? Track these meta-metrics:

The Iteration Scorecard

Product Metrics:

- Time to core value (decreasing?)

- Feature adoption rate (increasing?)
- User satisfaction scores (improving?)
- Support ticket volume (decreasing?)

Business Metrics:

- Conversion rate (improving?)
- Churn rate (decreasing?)
- Revenue per user (increasing?)
- Sales cycle length (shortening?)

Team Metrics:

- Iteration velocity (consistent?)
- Decision speed (quick?)
- Learning documentation (growing?)
- Team morale (high?)

The North Star Evolution

Your North Star metric should evolve with your iterations and as your business grows and moves through different stages:

BUSINESS STAGE	NORTH STAR METRIC (example)
Pre-product-market fit	No. weekly active users
Early growth	Month-over-month retention
Scaling phase	Revenue growth rate
Maturity	Market share defense

If your product iterations aren't moving your North Star, depending on your business stage, you're iterating on the wrong things.

Common Iteration Pitfalls and How to Avoid Them

Pitfall 1: The Everything Bagel

- Symptom: Product has every feature anyone ever requested
- Solution: For every feature added, remove one
- Mantra: "Perfection is achieved not when there is nothing left to add, but when there is nothing left to take away"

Pitfall 2: The Flip-Flop

- Symptom: Reversing iterations based on whoever spoke last
- Solution: Require data thresholds before reversing course
- Rule: No reversal without 2x the data that drove original change

Pitfall 3: The Local Maximum

- Symptom: Tiny optimizations while missing big opportunities
- Solution: Reserve 20% of iteration capacity for bold experiments
- Question: "What would a 10x improvement look like?"

Pitfall 4: The Iteration Theater

- Symptom: Lots of activity but no real progress
- Solution: Measure outcomes, not outputs
- Focus: Did user behavior actually change?

From Iteration to Innovation

The best iterations don't just improve your product—they teach you something fundamental about your market. Every iteration is a hypothesis test, and the learning compounds.

The Innovation Spiral

Start with small iterations, but let them guide you to bigger insights. For example:

Micro: "Users skip step 3 of onboarding"

Iterate: Remove step 3

Learn: Users don't care about feature X

Macro: Our value prop isn't what we thought

Micro: "Enterprise users keep asking for audit logs"

Iterate: Add basic audit logging

Learn: Compliance is a major purchase driver

Macro: We're actually a security company

The Pivot Decision Matrix

When do iterations add up to a pivot? It's time to pivot when you see:

- 3+ major iterations in the same direction
- Core assumptions repeatedly proven wrong
- Unexpected market pulling you strongly
- The original market is showing insufficient pain

But remember: pivots are just big iterations. The same discipline applies.

Your 30-Day Iteration Challenge

Theory without practice is useless. Here's your action plan:

Week 1: Instrumentation

- Set up analytics tracking
- Create a feedback collection system
- Define the North Star metric for your business stage
- Baseline current metrics

Week 2: First Iterations

- Identify the top 3 feedback patterns
- Score using the ICE framework
- Ship 2 small iterations
- Measure the immediate impact

Week 3: Acceleration

- Analyze week 2 results
- Ship 3-4 iterations
- Include one "wild card" experiment
- Start cohort analysis

Week 4: Rhythm Building

- Establish a weekly iteration ritual
- Document learning framework
- Plan next month's iterations
- Celebrate wins and learning

Daily Habit: End each day by asking: "What did we learn about our users today, and what will we do differently tomorrow?"

The Meta-Truth About Iteration

After years of iterating on products (and watching others do the same), here's what I know:

1. The founders who succeed aren't the ones with the best initial ideas. They're the ones who iterate fastest based on real feedback.
2. Your first idea is almost certainly wrong in important ways. That's not a flaw, that's the starting point.
3. Every iteration is a conversation with the market. The market always gets the last word.

4. The goal isn't to iterate forever. It's to iterate until you find the simple, powerful solution that makes previous iterations seem obvious in hindsight.

From Iteration to Conviction

You've identified real problems (Chapter 1), validated market demand (Chapter 2), understood your customers deeply (Chapter 3), found the solution for the problem (Chapter 4), and learned to iterate based on feedback (Chapter 5). Now comes the crucial transition: using iterations to find true problem-solution fit.

The next chapter will show you how to validate that your solution—refined through countless iterations—actually solves the problem in a way customers will pay for. It's where the rubber meets the road.

But first, start iterating. Today. Pick one piece of feedback you've been sitting on. Score it. Ship a small change. Measure what happens.

Because here's the truth: Reading about iteration is like reading about dancing. You can study the steps, understand the rhythm, memorize the movements. But you'll never learn to dance until you get on the floor and move.

Your market is playing the music. Time to start dancing.

6

Avoiding Common Validation Mistakes

The Graveyard of Good Intentions

I keep a folder on my laptop called **"The Graveyard."** It's packed with pitch decks, business plans, and post-mortems from startups that *should* have succeeded. Smart founders. Good ideas. Decent funding. Yet they all failed. Validation mistakes that seemed small and harmless at first proved to be fatal.

The cruel irony? Every one of these founders in that folder *thought* they were validating the right way. They ran surveys. They talked to customers. They built MVPs. They checked every box in the Lean Startup playbook. But they fell into traps so subtle that they didn't realize they were trapped until it was too late.

One trap can be especially deadly: **ignoring shifting market conditions.** While founders polished their product, interest rates spiked, supply chains snarled, or a new competitor rewrote the pricing game. Because they never tested whether their idea could survive those shocks, their once-promising business collapsed the moment reality moved the goalposts.

This chapter is about keeping you out of that graveyard. I'll show you the validation mistakes that kill startups—from focusing on flattering feedback to overlooking macro-economic curveballs—so you build something people want *and* something that can thrive when the market inevitably changes.

The Validation Theater Problem

Let's start with the most dangerous mistake: performing validation theater instead of actual validation. It looks like validation. It feels like validation. But it's actually an elaborate way to confirm what you already believe.

Signs You're Performing Validation Theater:

1. The Friendly Fire Problem. You're "validating" with:

- Your former colleagues who want you to succeed
- Friends who don't want to hurt your feelings
- Family members who don't understand the problem
- Network connections who are being polite

Real Validation: Talk to strangers who have zero incentive to make you feel good.

2. The Leading Question Trap. Your questions sound like:

- "Don't you hate it when [problem you're solving]?"
- "Wouldn't it be great if [your solution existed]?"
- "On a scale of 1-10, how much would this help you?"

Real Validation: "Tell me about the last time you dealt with [problem area]." Then shut up and listen.

3. The Solution Validation Fallacy. You're asking:

- "What features would you want?"
- "How much would you pay for this?"
- "Would you use this?"

Real Validation: Focus on the problem, not your solution. Validate that the problem exists, is painful, and is worth solving before discussing any solution.

The Confirmation Bias Machine

Our brains are wired to seek confirmation, not truth. Here's how it sabotages validation:

The Cherry-Picking Championship

- You interview 50 people
- 5 show strong interest
- 45 are lukewarm or negative
- You focus on the 5 and rationalize away the 45

The Interpretation Dance

- Customer: "That's interesting"
- Your brain: "They love it!"
- Reality: They're being polite
- Customer: "I might use something like that"
- Your brain: "They'll definitely buy!"
- Reality: They absolutely won't

The Memory Filter After a validation session, you remember:

- Every positive comment
- Vague interest as strong commitment
- Hypothetical scenarios as definite plans
- Your clever responses, not their hesitations

The Cure: Record everything. Review with a skeptical mindset. Have someone else review and point out the negative signals you're missing.

The Vanity Validation Metrics

Some validation metrics look impressive but mean nothing. Here's what to ignore:

Meaningless Metrics:

Survey Interest Scores "87% of respondents rated their interest as 7 or higher!"

- Reality: Interest without commitment is worthless
- Better metric: How many put down a deposit?

Feature Request Volume "Users requested 47 different features!"

- Reality: Feature requests often indicate core product confusion
- Better metric: What percentage achieve core value without requesting features?

Total Addressable Market Size "It's a $50 billion market!"

- Reality: TAM tells you nothing about problem severity or willingness to pay
- Better metric: How much are people currently spending to solve this problem?

Social Media Engagement "Our concept post got 500 likes!"

- Reality: Likes don't pay bills
- Better metric: How many commented asking how to buy?

Meaningful Validation Metrics:

The Credit Card Test How many people enter payment information for early access?

The Time Investment Test How much time will people invest in using your solution?

The Referral Test Do users recommend your solution without being asked?

The Retention Test Do people come back after the first use?

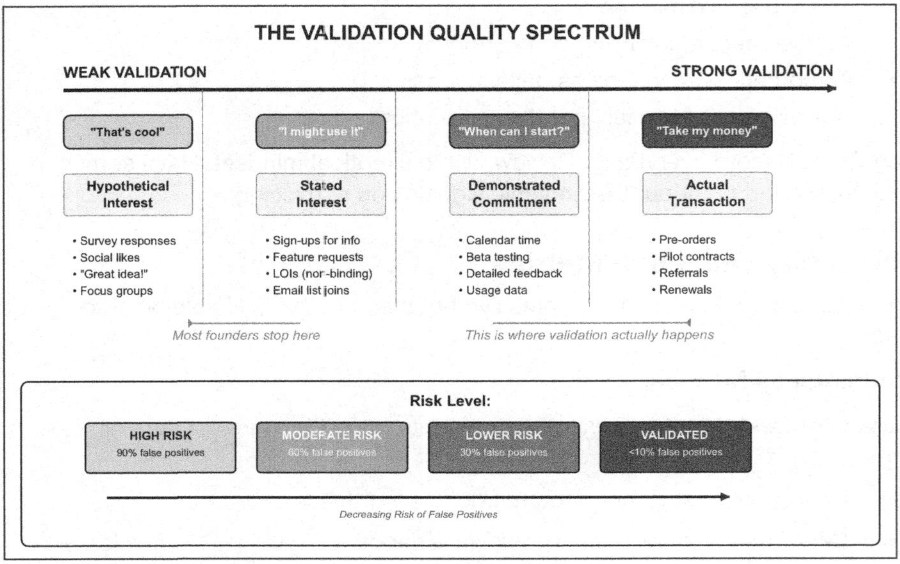

The Dangerous Validation Shortcuts

When pressure mounts—from investors, team members, or your own impatience—it's tempting to take shortcuts. Here are the most dangerous ones:

Shortcut 1: The Faster Horse Focus Group

You gather potential customers and ask what they want. They tell you faster horses. You build faster horses. You fail.

Why It Fails: Customers are experts at their problems, not solutions. They'll ask for incremental improvements to what exists, not breakthrough innovations.

Example: If Uber had asked people what they wanted in taxi services, they'd have said "cleaner cars" and "nicer drivers," not "push a button, get a ride."

The Right Way: Focus groups should explore problems and current behaviors, not solution ideas.

Shortcut 2: The Competitor Feature Checklist

You list all competitor features and validate that people want them. They say yes to everything. You build a bloated me-too product.

Why It Fails: People say yes to features in abstract but don't value them enough to switch from existing solutions.

Example: Every Slack competitor has "all of Slack's features plus more!" Yet Slack dominates. Features aren't the game.

The Right Way: Validate what jobs customers hire competitors for, then find a radically better way to do those specific jobs.

Shortcut 3: The MVP Excuse

"We'll just build it and see what happens. Fail fast, right?"

Why It Fails: Building even an MVP costs time and money. Failing fast is good. Failing unnecessarily is stupid.

Example: Color Labs raised $41 million and built a photo-sharing app without validating the specific problem. They shut down in a year.

The Right Way: Validate the problem and willingness to pay before writing a single line of code.

Shortcut 4: The Pitch Deck Validation

You create a beautiful pitch deck and investors get excited. You interpret fundraising interest as market validation.

Why It Fails: Investors bet on markets and teams, not validated solutions. Their excitement doesn't mean customers will buy.

Example: Juicero raised $120 million from top investors. Customers figured out they could squeeze the juice packs by hand. Company died.

The Right Way: Investor interest is a signal to investigate, not validation itself.

The Subtle Validation Mistakes

These are the mistakes that smart founders make because they're not obviously wrong:

Mistake 1: Validating with the Wrong Segment

You find a segment that loves your solution, but they're **not** your target market.

Example: Building a B2B tool and validating with freelancers who love it but can't afford enterprise pricing.

Signs You're Making This Mistake:

- Your most enthusiastic users can't afford your planned pricing
- The people who love it most have different workflows than your target
- You're getting pulled toward a market you don't want to serve

The Fix: Define your ideal customer profile first, then validate **exclusively** with that segment.

Mistake 2: The Time Horizon Mismatch

You're validating immediate pain but building for future behavior.

Example: "Once everyone works remotely, they'll need this!" (Built in 2015, died before 2020 proved them right)

Signs You're Making This Mistake:

- Your validation includes phrases like "once X happens"
- You're banking on behavior change that hasn't started
- Current solutions work "well enough" for now

The Fix: Validate problems that are painful **today**, not tomorrow.

Mistake 3: The Ecosystem Dependency

Your solution is validated, but depends on ecosystem changes outside your control.

Example: Building for a platform that might change its API, policies, or business model.

Signs You're Making This Mistake:

- Your success depends on another company's decisions
- You need multiple third parties to adopt your standard
- Regulatory changes are required for your model

The Fix: Validate not just demand but also your ability to deliver **independently**.

Mistake 4: The Validation Sample Size Delusion

You achieve statistical significance without practical significance.

Example: "500 survey respondents showed interest!" But you need 5 million users to build a viable business.

Signs You're Making This Mistake:

- Your validation math requires capturing an unrealistic market percentage
- You're extrapolating from a tiny sample to a huge market
- Geographic or demographic limitations in your validation

The Fix: Validate the **density** of demand, not just the existence of demand.

Mistake 5: The Static-Market Assumption

You validate in today's climate, then build as if that climate will never change.

Example: Launching a discretionary-spend consumer app during a boom year, only to watch customers vanish when interest rates spike and wallets tighten.

Signs You're Making This Mistake

- Rising acquisition costs or churn traced to macro-economic shifts
- Key customers are slashing budgets while you assume constant spend
- Competitors cutting prices or pivoting—yet your plan stays frozen
- No scenario analysis or "what-if" questions in your validation notes

The Fix: Stress-test your idea against multiple market conditions.

1. **Run "recession," "boom," and "status quo" scenarios** for pricing, demand, and cash burn.
2. **Re-validate quarterly**—short interviews or surveys to gauge whether willingness-to-pay or usage is shifting.
3. **Track lead indicators** (interest rates, supply-chain costs, regulatory chatter) and set trigger points for a pivot.
4. **Build flexibility** into pricing, cost structure, and product roadmap so the business can bend without breaking when the market shifts.

Spot these five mistakes early, and you'll validate what customers want **and** what the market will still reward when the winds inevitably change.

The Emotional Validation Traps

These mistakes happen because we're human:

The Sunk Cost Validation

You've already built something, so you find ways to validate it retroactively.

Internal Dialog: "We've spent six months on this. There must be a market. Let's keep looking."

The Reality Check: Time invested doesn't create market demand. Kill zombies quickly.

The Founder Autobiography Validation

The problem you're solving is so personal that you can't see beyond your own experience.

Internal Dialog: "I had this problem intensely, so others must too."

The Reality Check: Your experience might be unique. Validate that others share both the problem AND your willingness to pay for solutions.

The Team Harmony Validation

You avoid hard validation questions because they might reveal uncomfortable truths.

Internal Dialog: "The team is excited and working hard. Let's not kill morale with negative feedback."

The Reality Check: Better to face hard truths early than enthusiastically build the wrong thing.

The Pre-Mortem Exercise

Before you declare your idea validated, run this pre-mortem:

Imagine it's two years from now and your startup has failed. Working backward:

1. What validation signal did you misinterpret?
2. What assumption seemed reasonable but proved false?
3. What customer segment did you think would buy but didn't?
4. What competitor or alternative did you underestimate?
5. What behavior change did you expect that didn't happen?

If any of these feel plausible, you have more validation work to do.

The Validation Decision Framework

After all your validation efforts, you need to make a decision. Here's a framework:

The Validation Decision Framework
Green Light Criteria :: All of these must be true
1. Identified a problem that causes measurable pain
2. Found people currently paying to solve it
3. Discovered why current solutions fail
4. Validated willingness to pay your price
5. Confirmed you can reach customers affordably
6. Verified no deal-breaking dependencies
Yellow Light Criteria :: Proceed with caution if any of these are present
1. Some validation signals are mixed
2. Market timing might be slightly early
3. Competitive landscape is heating up
4. Unit economics are borderline
Red Light Criteria :: Any one of these means STOP
1. People say they want it but won't pay
2. The problem is real but not painful enough
3. Existing solutions are "good enough"
4. Customer acquisition cost exceeds lifetime value
5. You're validating a vitamin, not a painkiller

Your Validation Audit Checklist

Run through this checklist honestly:

The Validation Audit Checklist
Problem Validation
• Validated with 20+ target customers
• Found evidence of current spending
• Identified specific, measurable pain
• Confirmed problem frequency
• Discovered emotional impact
Solution Validation
• Tested with real prototype/demo
• Measured actual behavior, not opinions
• Achieved 40%+ "very disappointed" without product
• Got unprompted referrals
• Saw organic pull for features
Economic Validation
• Tested specific price points
• Calculated realistic CAC
• Modeled believable LTV
• Found scalable acquisition channels
• Validated sales cycle length
Competition Validation
• Understood why incumbents vulnerable
• Found defensible differentiation
• Validated switching triggers
• Confirmed distribution advantages
• Tested unique value prop
Score
16-20 checked = Strong Validation
11-15 checked = Needs Work
Below 10 = Back to Discovery

The Meta-Lesson About Validation

After watching hundreds of startups succeed and fail, here's what I know:

Validation isn't about proving your idea right. It's about discovering what's true.

The goal isn't to feel confident. It's to be correct.

Every founder who ended up in my "Graveyard" folder felt validated. The ones who succeeded questioned their validation constantly.

The market doesn't care about your validation process. It only cares whether you've built something it wants.

From Validation to Victory

You've learned to identify real problems (Chapter 1), research your market (Chapter 2), talk to customers (Chapter 3), iterate based on feedback (Chapter 4), and avoid validation traps (Chapter 5).

The next chapter brings it all together: building validation into your company's DNA so that customer discovery becomes a continuous competitive advantage, not just a one-time exercise.

But first, audit your current validation. Where are you taking shortcuts? What uncomfortable truths are you avoiding? Which metrics are you using for comfort rather than clarity?

Because here's the thing about validation mistakes: they're only expensive if you keep making them. The moment you spot them and correct course, they become valuable lessons.

Your customers are trying to tell you the truth. The question is, are you brave enough to hear it?

The startups that succeed aren't the ones that avoid validation mistakes entirely. They're the ones that catch them early, admit them honestly, and correct them quickly.

Time to be one of those startups.

7

Validation Case Studies & Quick Wins

The Proof Is in the Pattern

After years of working with startups, I've noticed something fascinating: successful validation stories follow remarkably similar patterns. Not in the details, every market is unique, but in the approach, the discipline, and the willingness to face hard truths.

This chapter showcases real validation journeys that led to breakthrough successes. But more importantly, I'll extract the repeatable patterns you can apply to your own startup. These aren't just feel-good success stories. They're blueprints for validation done right.

Plus, I'll share some quick wins and validation techniques you can implement this week that deliver significant insights with minimal effort. Because sometimes you need momentum as much as methodology.

Case Study 1: Buffer - The Landing Page That Built a Business

The Context

Joel Gascoigne had an idea for a tool to schedule social-media posts, but didn't want to build anything without first validating demand. This was 2010, years before "lean startup" language became ubiquitous.

The Validation Journey

Step 1: The Two-Page MVP

Joel created a simple two-page website:

- Page 1: Described the product and its benefits
- Page 2: Listed pricing plans for a product that didn't yet exist

When visitors clicked “Sign up,” they saw a message along the lines of: “You caught us before we’re ready. Please leave your email, and we'll notify you when we launch."

Step 2: The Price Test

After getting enough emails showing interest, Joel added a crucial third page: a dedicated pricing page. The revised flow was:

- Page 1: Product description
- Page 2: Pricing plans
- Page 3: Email collection

This step tested whether people were willing to pay the proposed price, not just whether they liked the idea.

Step 3: The Manual MVP

With validated interest and price acceptance, Joel built a minimal version. But here's the kicker—the first version barely automated anything. Joel manually scheduled tweets for early users to confirm that the core value proposition worked.

The Results

- Hundreds of email sign-ups validated initial interest.
- Around half of visitors who reached the pricing page proceeded onward, indicating that the price point was acceptable.
- Joel’s first paying customer arrived within 4 days of launching the minimal product, with a small one-time payment that he described as “jumping-around-the-room” exciting.
- Buffer grew to **$5 million+ ARR within about 3 years.**

The Pattern to Copy

1. Test interest before building anything
2. Test pricing before building anything
3. Deliver value manually before automating
4. Let customer behavior guide feature development

The Quick Win

Create a "coming soon" landing page this week. Drive 100 targeted visitors to it. If fewer than 10% give their email, your value proposition needs refinement.

For founders, Buffer’s origin is a lesson that **you don’t need code to validate pricing or demand; a landing page, a clear value proposition, and a willingness to manually deliver the outcome are enough to find out whether anyone is willing to pay before you build a full-scale product.**

Case Study 2:
Pebble - Crowdfunding as Validation

The Context

Eric Migicovsky had built a smartwatch prototype, but traditional investors were hesitant. Rather than keep pitching a market that many people still doubted, he turned to Kickstarter to test demand directly with customers.

The Validation Journey

Step 1: The Prototype Proof

Pebble showed a working prototype, which helped backers understand what the product would actually do.

Step 2: Direct Demand Testing

The Kickstarter campaign let the team reach early adopters directly and see whether enough people cared about the product to back it.

Step 3: Price Validation

The campaign used a $99 pledge level, and demand was strong enough that Pebble eventually capped smartwatch availability at 85,000 units.

The Results

Pebble reached its $100,000 goal in about two hours and went on to raise **$10,266,845** from **68,929 backers.**

The Pattern to Copy

1. Build a prototype that proves the core idea.
2. Put it in front of early adopters.
3. Use pre-orders or crowdfunding to validate demand before manufacturing at scale.

The Quick Win: Create a pre-order campaign (even without crowdfunding). If you can't get 100 people to pre-order at 50% off, you don't have product-market fit.

For founders, Pebble's story is simple: If you can't convince VCs, convince customers first. A prototype plus a pre-order test is the most honest way to find out whether your product is worth building at all.

Case Study 3:
Dropbox - The Video That Validated a Market

The Context

Drew Houston faced a chicken-and-egg problem. Dropbox needed robust infrastructure to sync files seamlessly across devices, yet building it all up front would be expensive. He decided to validate demand before investing millions in development.

The Validation Journey

Step 1: The Explainer Video

Houston created a simple 3-minute screencast showing how Dropbox would work. He didn't just list features; he showed the product solving a real-world problem—dragging a file into a folder and seeing it appear instantly on another device—which made the value concrete and visceral.

Step 2: The Community Seeding

Instead of broad marketing, he posted the video on Hacker News and Digg, platforms dense with tech-savvy early adopters who felt the file-sync problem acutely. This targeted distribution amplified the signal and ensured the audience understood the problem deeply.

Step 3: The Waiting List Hack

The video drove viewers to a landing page to sign up for a private-beta waiting list. The list grew from roughly **5,000 to 75,000 overnight**, without a finished product yet. The trial-waitlist mechanic itself validated both demand and the product's viral potential.

The Results

- The beta waiting list surged from about **5,000 to 75,000 users in a single day**, all from a video-based "MVP" rather than a shipping product.
- No full-scale product existed at this stage; the explainer video plus waitlist provided enough proof of market demand.
- Houston used this validated traction to secure Y Combinator admission and later raise funding.

The Pattern to Copy

1. Show, don't tell: demonstrate value in a realistic workflow, not a feature list.
2. Go where your early adopters already congregate, and speak to their pain directly.
3. Bake viral mechanics into your validation (e.g., waitlists, referrals, or sharing incentives).
4. Use concrete validation metrics (sign-ups, wait-list growth) when pitching investors.

The Quick Win

Create a 2-minute video showing your product solving a real-world problem. Share it in 3 relevant communities where your target users hang out. If you don't get at least 100 upvotes, likes, or shares combined, your positioning or problem framing needs refinement.

For founders, Dropbox's story is a lesson that you don't need to ship software to validate a market; a **simple, realistic video that shows the product in use can generate enough traction and data to prove demand and justify the investment in building the real product.**

Case Study 4: Airbnb - The Concierge MVP

The Context

Brian Chesky and Joe Gebbia wanted to validate whether strangers would really stay in each other's homes. The trust barrier seemed huge, especially in 2007–2008, before the platform existed.

The Validation Journey

Step 1: The Event Trigger

They didn't start with a "global sharing economy" pitch. Instead, they focused on a specific event at the **Industrial Design conference in San Francisco**, when local hotels were sold out and travelers desperately needed places to stays.

Step 2: The Manual Everything

The first Airbnb was effectively a concierge service, not a platform:

- A simple website describing "AirBed & Breakfast."
- Air mattresses in their own apartment.
- Breakfast offered to guests.
- Manual payments and personal hosting by the founders.

Step 3: The Geographic Expansion

After validating the idea in San Francisco, they started expanding to other cities where hotel shortages and events created similar pain points. At first, they added listings manually, often personally photographing and supporting early hosts to understand the real-world friction.

The Results

- Their very first test had **three guests** during the design-conference event, proving strangers were willing to sleep in someone's apartment.
- Early expansion was careful and city-by-city; they didn't build a full-scale platform until they saw traction and understood how the model worked.

- They achieved reasonable early-stage **unit economics through manual operations** (e.g., carefully curating listings and photos, then iterating based on what increased bookings) before heavy automation.

The Pattern to Copy

1. Start with **acute, specific pain** (a sold-out event in one city) rather than a vague "everyone everywhere" market.
2. Do things that don't scale: **manual payments, manual onboarding, manual hosting** to validate core assumptions.
3. Expand **market by market**, not globally, and learn how different cities behave.
4. Achieve positive **unit economics manually** before pouring resources into technology.

The Quick Win

Manually deliver your service to **5 customers this week**. If you can't make the economics and experience work manually, automation won't save you.

For founders, Airbnb's origin is a reminder that the best way to test an "impossible" market isn't a perfect product, but a **small, specific pain point faced by real people—and the courage to manually deliver the solution yourself before you build anything at scale**

Case Study 5: Zappos - The Inventory-Free Validation

The Context

Nick Swinmurn wanted to test whether people would buy shoes online, a seemingly risky idea at the time because "you can't try on shoes" was the conventional wisdom. Rather than betting millions on inventory and warehousing, he designed a low-risk validation playbook.

The Validation Journey

Step 1: The Wizard of Oz Backend

Swinmurn went to local shoe stores, took photos of their inventory, and posted them on a simple website (initially under the name "Shoesite"). When someone "ordered" online, he would:

- Buy the shoes at full retail price from the store.
- Package them and ship them himself.

There was **no own inventory, no supplier contracts, and no warehouse**—just a manual, behind-the-scenes operation that validated whether people would pay to buy shoes sight-unseen.

Step 2: The Return Policy Innovation

To overcome the "can't try on" objection, Zappos offered **free returns**, often with pre-paid shipping labels. This wasn't just a convenient perk; it was a test of whether the **try-on objection collapsed when customers bore no risk**. The experiment showed that free returns dramatically reduced friction and made the model viable.

Step 3: The Service Differentiation

As Zappos scaled, the team kept testing increasingly generous service policies (e.g., extended return windows, 24/7 support, and surprise upgrades) to see what drove loyalty and word-of-mouth. The data showed that **customer service and reliability became bigger differentiators than selection alone.**

The Results

- The **first few online sales** proved that people would, in fact, buy shoes without trying them on, validating the core hypothesis.
- The **return rate eventually settled around 10%**, which the company treated as evidence that the model was workable, not catastrophic.
- Zappos built a **repeat-purchase-driven business**; founders and later Tony Hsieh frequently cite strong repeat-purchase behavior as proof that the customer-service bets were paying off. Your "75% repeat-purchase rate" is directionally in line with the common narrative, but exact numbers are rarely published, so it's safer to keep that phrasing generic (e.g., "high repeat-purchase rate," "strong repeat-purchase behavior").
- In 2009, Zappos was acquired by Amazon for **$1.2 billion**, the result of years of validation-driven iteration starting from a nearly inventory-free setup.

The Pattern to Copy

1. **Remove the inventory barrier to validation:** don't buy stock before you know people will pay.
2. **Test objection-handling early:** free returns, risk-free trials, or guarantees reveal whether the "real" objection is real.
3. **Validate differentiators beyond the core product:** test generous policies and service levels to see what drives loyalty.
4. **Let unit economics guide expansion:** once you can make the manual, low-inventory model work profitably, scale infrastructure.

The Quick Win

Sell something you don't own yet. Take orders, then source. If you can't make **10 sales this way**, adding inventory and infrastructure won't save you. This is your true "zero-inventory" validation test.

For founders, Zappos' origin is a master class in risk-free validation: **Instead of gambling on inventory or branding, test the core assumption first — can people be trusted to buy without touching? Then design guarantees and service so good that the old "can't try on" excuse stops mattering.**

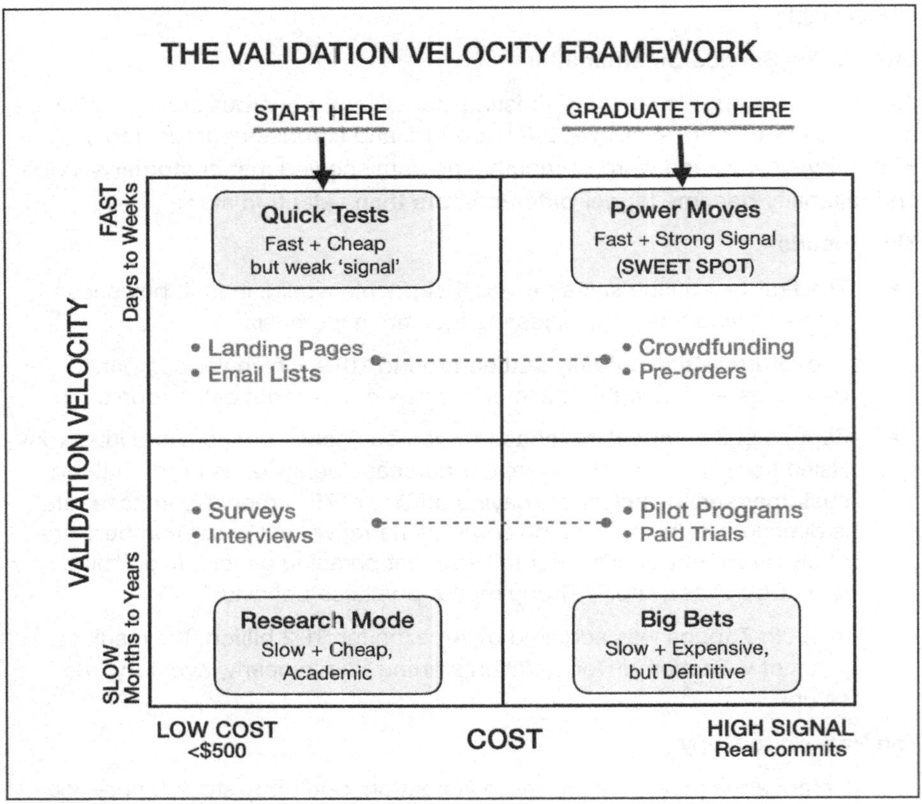

High Validation Velocity = Startup Survival Advantage

In the startup battlefield, **validation velocity** — the speed and frequency with which you test and kill bad assumptions — is the difference between thriving and dying broke. The framework you've diagrammed brilliantly captures this: founders who master **Quick Tests** (landing pages, email lists) iterate weekly, quickly separating signal from noise, while those stuck in **Research Mode** (surveys, academic studies) burn cash on low-signal insights. As Buffer, Dropbox, and Airbnb prove, high-velocity founders graduate to **Power Moves** like pre-orders and pilots only after cheap tests confirm the path, achieving **sweet-spot validation** that combines speed with real commitment—ultimately compressing years of uncertainty into months of momentum.

The Quick Win Playbook

Here are validation techniques you can implement immediately:

Quick Win 1: The Mom Test Email

Send this email to 20 potential customers:

"Hi [Name],

I'm researching how [job title] handles [problem area]. Not selling anything—just trying to learn.

Would you have 15 minutes this week to share how you currently deal with [specific task]?

Happy to share what I learn with you.

[Your name]"

Success metric: 5+ responses willing to talk

Quick Win 2: The Fake Feature Test

Add a button for a feature that doesn't exist to your landing page or product. When clicked, show: "Coming soon! Want early access?" Track:

- Click rate (tests interest)
- Email submission rate (tests real interest)
- Follow-up engagement (tests urgency)

Success metric: 10%+ click rate, 50%+ email submission

Quick Win 3: The Price Anchor Test

Create three pricing tiers:

- Budget: What you think people will pay
- Standard: 2x the budget price
- Premium: 5x the budget price

Present to potential customers and ask which they'd choose and why.

Success metric: 20%+ choose Standard or Premium

Quick Win 4: The Competitor Customer Interview

Find 5 people using competitor products. Ask:

- "What made you choose [competitor]?"
- "What's the most frustrating part about it?"
- "What would make you switch?"

Success metric: Consistent switching triggers identified

Quick Win 5: The LinkedIn Message Test

Send this to 50 target customers on LinkedIn:

"Hi [Name], I noticed you're a [job title] at [company type]. I'm researching how teams like yours handle [problem]. Would you pay for a solution that [specific benefit]? Just validating an idea—would love your quick thoughts."

Success metric: 10+ substantive responses

The Validation Patterns That Matter

Across all these case studies, certain patterns emerge:

Pattern 1: Specificity Beats Generality

- Airbnb: Conference attendees, not all travelers
- Pebble: Tech enthusiasts, not all watch wearers
- Buffer: Social media managers, not all marketers

Pattern 2: Behavior Beats Intention

- Money committed > Email submitted > Survey response
- Time invested > Opinion shared
- Action taken > Promise made

Pattern 3: Manual Before Automated

- Validate value delivery before scaling
- Human insight before algorithmic optimization
- High touch before high tech

Pattern 4: Iteration Based on Data

- Each validation creates the next hypothesis
- Customer behavior drives feature priorities
- Market feedback trumps founder vision

Your 7-Day Validation Sprint

Stop reading. Start validating. Here's your plan:

Day 1: Choose one quick win and implement it

Day 2: Analyze results and pick the next test

Day 3: Create a simple landing page or prototype

Day 4: Drive 50 targeted visitors to test

Day 5: Conduct 5 problem validation interviews

Day 6: Analyze all data and identify patterns

Day 7: Decide: proceed, pivot, or probe deeper

Success Criteria:

- One clear signal of demand
- Three specific learnings about customers
- Next validation step identified

The Meta-Truth About Validation

After studying hundreds of validation journeys, here's what separates the successes from the failures:

Successful validators fall in love with the problem, not their solution.

They measure what people do, not what they say.

They start small and specific, then expand based on evidence.

They treat validation as the beginning of customer development, not the end.

Most importantly, they have the courage to kill ideas that don't validate, no matter how much they love them.

From Validation to Execution

You've now seen how real companies validated their way to success. You have a playbook of quick wins. You understand the patterns that matter.

The final chapter will show you how to build validation into your company's ongoing DNA—because customer discovery isn't a phase, it's a philosophy.

But first, pick a quick win. Run it this week. Not next week. Not when you're ready. This week.

Because here's what I've learned from every successful validation story: The founders who succeed don't wait for perfect conditions. They start where they are, with what they have, and let customer truth guide them forward.

Your customers are waiting to teach you everything you need to know. The only question is: Are you ready to learn?

The best validation starts with a single conversation, a simple test, a small bet.

Make yours today.

Your Next Steps

Your Customer Discovery Journey Starts Now

We've covered a lot of ground together. From understanding market pain to mastering customer interviews, from avoiding validation traps to learning from successful case studies. But here's the thing about customer discovery: reading about it is like studying a map. Useful, yes. The real learning happens when you start walking the territory.

The startups that succeed aren't the ones with the best ideas or the most funding. They're the ones that get closest to their customers, fastest. They're the ones that replace assumptions with insights, hunches with data, and ego with empathy.

The Three Truths That Matter

After all we've covered, everything boils down to three fundamental truths:

Truth #1: Your first idea is wrong. Not *completely* wrong, perhaps, but wrong in ways that matter. The founders who succeed are the ones who discover exactly how they're wrong and fix it before they run out of money.

Truth #2: Customers can't tell you what to build. But they can tell you everything about their problems, their current solutions, and what success looks like. Your job is to listen to their problems, not their solutions.

Truth #3: Validation is never done. Markets change. Customers evolve. Competitors adapt. The day you stop discovering is the day you start dying.

Your 30-Day Action Plan

Knowledge without action is worthless. Here's what to do next:

Week 1: Problem Discovery

- Interview 10 potential customers about their problems (not your solution)
- Map out their current workflows and pain points
- Identify patterns and prioritize the most acute problems

Week 2: Market Research

- Size your market using three different methods
- Analyze 5 competitors and their customer reviews
- Define your beachhead market with surgical precision

Week 3: Solution Validation

- Create a simple landing page or prototype
- Test with 20 target customers
- Measure behavior, not opinions

Week 4: Iteration and Decision

- Synthesize all learnings into key insights
- Make the hard decision: proceed, pivot, or probe deeper
- Plan your next 30-day sprint based on what you learned

The Resources to Keep You Going

Customer discovery is a journey, not a destination. To continue your education:

- **Revisit Chapter 3** whenever you're preparing for customer interviews
- **Use the frameworks from Chapter 2** for ongoing market research
- **Check your validation against Chapter 5's warnings** before major decisions
- **Return to Chapter 1** whenever you lose sight of the problem you're solving

For broader startup guidance, my comprehensive guide "How to Start Your Start-Up" covers everything from incorporation to investment, team building to scaling.

The Most Important Step

But forget all that for a moment. The most important thing you can do is this:

Talk to one potential customer today.

Not tomorrow. Not next week. Today.

Send that email. Make that call. Schedule that coffee. Because every day you delay is a day you're building on assumptions instead of insights.

Your Competitive Advantage Awaits

In a world where most founders fall in love with their solutions, your willingness to fall in love with your customers' problems is a superpower. In a market where everyone's building first and asking questions later, your discipline to validate before you build is a competitive advantage.

Customer discovery isn't just a phase you go through. It's a mindset you adopt. It's the difference between building something you hope people want and building something you know they need.

The Final Challenge

I'll leave you with a challenge that's served me well over the years:

For the next 30 days, end every day by asking yourself: **"What did I learn about my customers today that I didn't know yesterday?"**

If you can't answer that question, you spent the day building on assumptions.

If you can answer it, you're one step closer to building something that matters.

Now Go Discover

The market is waiting to teach you everything you need to know. Your customers are ready to show you the way. The only thing standing between you and a successful startup is your willingness to listen, learn, and iterate.

You have the tools. You have the knowledge. You have the frameworks.

Now go out there and discover what your customers really need.

Because, in the end, startups don't fail solely from poor execution or a lack of funding.

They fail by building things nobody wants.

You're not going to be one of them.

Your customer discovery journey starts now.

Go make it count.

Remember: Every unicorn started with a founder who was brave enough to admit they didn't have all the answers—and curious enough to go find them.

Made in the USA
Monee, IL
04 November 2026

67379481R00066